THE DISTRACTION ADDICTION

THE DISTRACTION ADDICTION

Getting the Information You Need and the Communication You Want Without Enraging Your Family, Annoying Your Colleagues, and Destroying Your Soul

ALEX SOOJUNG-KIM PANG

LITTLE, BROWN AND COMPANY
New York Boston London

Little, Brown and Company
Hachette Book Group
237 Park Avenue, New York, NY 10017
littlebrown.com

First Edition: August 2013

ISBN 978-0-316-20826-0 (HC); 978-0-316-24752-8 (int'l. ed.)
LCCN 2013935729

10 9 8 7 6 5 4 3 2 1

RRD-C

Printed in the United States of America

For Heather

CONTENTS

ENTRANCE TO THE SAND WALK.

The Sandwalk, the walking path installed by Charles Darwin at Down House, from a 1929 photograph (© The British Library Board, 010822.de.81)

THE DISTRACTION ADDICTION

INTRODUCTION

TWO MONKEYS

On the western edge of the ancient city of Kyoto, Japan, on the slope of Mount Arashiyama (literally "Stormy Mountain"), stands the Iwatayama Monkey Park. The park has winding paths and fine views of Kyoto, but the main attraction is the tribe of about a hundred and forty macaques who live there. The monkeys of Iwatayama are famously gregarious, playful, and, occasionally, crafty. Like all members of the *Macaca* genus, they combine sociability and intelligence. They play with their kin, watch one another's young, learn new skills from one another, and even have distinctive group habits.

Some develop a mania for bathing, snowball-making, washing food, fishing, or using seawater as a seasoning. Iwatayama macaques are known for flossing and for playing with stones. This has led some scientists to argue that macaques have a culture, something we've traditionally thought of as distinctly human. They're also humanlike in their natural curiosity and cunning: one second, you're watching one do something cute, and the next second, his friends are making off with the bag of food you bought at the park's entrance.

They're like humans in one other way. For all their smarts, nothing keeps their attention for very long. The mountainside gives them a fantastic view of one of the world's most historic cities, but it doesn't impress them. They keep up a constant chatter, a running monologue of inconsequence. The macaques are living examples of the Buddhist concept of the monkey mind, one of my favorite metaphors for the everyday, undisciplined, jittery mind. As Tibetan Buddhist teacher Chögyam Trungpa explains, the monkey mind is crazy: It "leaps about and never stays in one place. It is completely restless."

The monkey mind's constant activity reflects a deep restlessness: monkeys can't sit still because their minds never stop. Likewise, most of the time, the human mind delivers up a constant stream of consciousness. Even in quiet moments, minds are prone to wandering. Add a constant buzz of electronics, the flash of a new message landing in your in-box, the ping of voicemail, and your mind is as manic as a monkey after a triple espresso. The monkey mind is attracted to today's infinite and ever-changing buffet of information choices and devices. It thrives on overload, is drawn to shiny and blinky things, and doesn't distinguish between good and bad technologies or choices.

The concept of the monkey mind appears throughout Buddhist teachings—one small indicator of the fact that the mind and its relationship to the world have been studied deeply for thousands of years. Every religion has contemplative practices, calls to use silence and solitude to quiet the mind. In John Drury's introductory note to the Anglican *Matins and Evensong,* he exhorts worshippers "to be patient and relaxed enough to allow a long tradition to have its say" and "allow our own thoughts and feelings to become closer to us than life outside admits." Only then can one fully enter "the cool and ancient order of the services which gives a space and a frame, as well as cues, for reflections on our regrets and hopes and gratitudes." Catholic monastics treat meditation as preparing the mind to receive God's wisdom; the busy mind cannot hear the divine. In Buddhism, though, mental discipline is more an end in itself, rather than just a means to an end. The everyday mind is like churning water; learn to make it still, like the mirror-flat surface of a calm lake, Buddhists say, and its reflection will show you everything.

A few miles away from Iwatayama, a robotics laboratory at Kyoto University houses a robot controlled by another monkey, a rhesus named Idoya. Incredibly, Idoya isn't in Japan; she lives in North Carolina, in a neuroscience laboratory at Duke University, and her brain is connected to the robot via the Internet. The laboratory is run by neuroscientist Miguel Nicolelis, who, to make things just a bit more global, was born and educated in Brazil. Nicolelis has been studying the brain and how the brain changes as it learns executive functions; he's also developed a specialty in what scientists call brain-computer interface (BCI) technologies. Today you can buy primitive brain-wave readers that can control video games, and scientists are mapping brain functions and testing the brain's ability to control complex objects through BCIs. Eventually, they hope, BCIs will be used to route brain signals around damaged nerves, restoring body control to people with spinal-cord injuries or neurodegenerative disorders.

Idoya is the latest in a series of monkeys Nicolelis has worked with. Over the previous decade, he and his team demonstrated that a monkey with electrodes implanted in its brain could operate joysticks or robotic arms with its mind. Brain scans showed something remarkable: the neurons in the monkey's frontoparietal lobe—the section that controlled the animal's arms—fired when the monkey operated a robot arm. In other words, the monkey's brain stopped treating the robot arm as a tool, as something that it used but that was clearly separate from itself. The brain remapped its picture of the monkey's body to incorporate the robot arm. At the neural level, the distinction between the monkey's arms and the robot arm blurred. As far as the monkey's brain was concerned, monkey arms and robot arm were all part of the same body. Nicolelis and his colleagues in Japan implanted electrodes in the section of Idoya's brain that regulated walking; they then taught her to walk on a treadmill and studied how her brain's neurons fired as she walked. When she obeyed commands to speed up or slow down, she was rewarded with food. They then put a video monitor in front of the treadmill. Instead of showing *The View* or CNBC, however, the screen showed Idoya a live video feed of CB-1, the human-size robot in Kyoto. (CB-1 itself is a prodigy. Equipped with four video cameras, gyroscopic stabilizers, and hands that can grasp objects, it can hold a bat, swing at baseballs, and learn manual tasks by imitating humans.)

When Idoya started walking while looking at the Kyoto robot on the monitor, the electrodes in her brain picked up the signals generated by the neurons that control locomotion. The signals were transmitted over the Internet to CB-1, which, following those same signals, stepped with her. The better she controlled the robot, the more treats she got. After an hour of Idoya walking and munching Cheerios, the scientists switched off her treadmill. Still focused on the screen, the monkey stopped walking—but she kept CB-1 on course, and for the next several minutes, she continued to make it walk. Once again, Nicolelis's team had shown that a primate brain could learn to directly control robots—and, in the process, that brain would start to treat the robot as an extension of its own body. Brain scans showed that Idoya's brain performed exactly the same way whether she was using her own flesh-and-fur legs or the electronic-and-plastic ones. As far as her brain was concerned, there was no longer any difference between the two.

Idoya and the monkeys of Iwatayama represent two different sides of the human mind, two contrasting relationships with information technology, and two futures. The chattering monkey is the untutored, undisciplined reactive mind, the mind that loves stimulation but doesn't hold a thought. The cyborg monkey represents a mind that isn't overwhelmed by technology, because it no longer experiences the technologies it uses as separate from itself, as requiring conscious effort and attention. A mix of deliberate practice, tinkering and experimenting, and neural rewiring have created an extended mind in which brain, body, and tools are entangled and work together effortlessly.

For too long, we've left the chattering monkey in charge of our technologies, and then we wonder why things go bad. We want to be like the cyborg monkey (albeit not as hairy and without the electrodes). We want that same capability to use complicated technologies without thinking about them, without experiencing them as burdens and distractions. We want our technologies to extend our minds and augment our abilities, not break up our minds.

Such control is within our reach. Rather than being forced into a state of perpetual distraction, with all the unhappiness and discontent such a state creates, we can approach information technologies in a way that is mindful and nearly effortless and that contributes to our ability to focus, be creative, and be happy.

It's an approach I call contemplative computing.

The term sounds oxymoronic. What could be less contemplative than today's technology-intensive environment? What could possibly be less conducive to a clear, meditative state than interactions with computers, cell phones, Facebook, and Twitter?

Contemplative computing isn't enabled by a technological breakthrough or scientific discovery. You don't buy it. You do it. It's based on a blend of new science and philosophy, some very old techniques for managing your attention and mind, and a lot of experience with how people use (or are used by) information technologies. It shows you how your mind and body interact with computers and how your attention and creativity are influenced by technology. It gives you the tools to redesign your relationships with devices and the Internet, to make them work better for you. It's a promise that you can construct a healthier, more balanced relationship with information technology.

To get a sense of how that can happen, let's look at what digital life is like for many of us—and then at what it could be like.

Imagine a Monday morning. You reach over to the nightstand, grab your smartphone, and switch off the alarm. You rub your eyes with one hand and open the phone's e-mail program with the other. You're not really awake; you do it automatically. You watch the icon spin as the phone connects to your e-mail server.

Nineteen messages in your in-box. Most are automatically generated newsletters, coupons, daily deals, or social-media updates; six are from colleagues up even earlier than you. You answer one, start another, and realize you're not sure what to say, so you flip over to the Web browser and check the news. You'll finish the message later. European bankers arguing over the terms of the latest bailout . . . another NASDAQ flash crash . . . roundup of blog posts commenting on the suicide of a cast member of a reality-TV show. . . . Suddenly you realize it's been twenty minutes. Gotta get up.

On the train to work, you look out the window and see a driver holding his phone and steering wheel in the same hand as he uses the phone to navigate, and another person steering with one hand and texting with the other. It makes driving while talking on your cell phone seem downright cautious. The police should give out more tickets to distracted drivers, you think, but as more cruisers are equipped with laptops, more police are getting distracted too.

Work turns out to be one of those days: these coworkers need numbers, those colleagues need your feedback; can you help with this problem, explain these options, talk to this person? It's one thing when there's lots of input directed toward one goal, but this multitasking is something else entirely. You're used to dealing with a constant stream of interruptions, but today, even your interruptions get interrupted. It's hard to say no, and it's hard to get back on task. After each interruption, you need a couple minutes to remember what you were doing, to gather your thoughts and start again.

By late afternoon, you're finally ready to print out your work. You hit Print, and an error message appears: You need to update your printer driver. When you click Okay, a minute goes by, and then there's another message: The latest driver isn't compatible with the old version of your operating system. You or your IT department need to update that too. Half an hour later, you restart your computer and finally print out your work. The experience is frustrating, but it's not at all unusual. According to a 2010 Harris interactive poll (sponsored by tech giant Intel), computer users spend an average of forty-three minutes every day—five hours a week, or eleven days a year—waiting for computers to start up, shut down, load software, open files, connect to the Internet.

On your way to meet a friend for a drink after work, you pass by people who are focused on their cell phones and have trouble tearing their attention away from their screens. You feel your phone buzzing in your pants pocket, but when you reach to answer it, there's no phone there. You check your other pockets, worried that you've lost it. The last time it happened, you felt like part of your brain had been shut off. But it's a false alarm: it's in your jacket.

During drinks, you and your friend each get the occasional text. Conversation flows and trails off as you each look at your phone, nearly finish a thought as you start typing. One message from an old ex is especially weird: It's all garbled, and it's the middle of the night in that time zone. "I've heard of that happening," your friend says, not looking up from her phone. "She's probably sleep-texting." Really? "It's like sleepwalking"—*type-type-type*—"except you"—*type-type-type*—"you know, text people."

It makes sense that some of us would start texting in our sleep. After all, information technologies and the Internet have thoroughly insinuated themselves into our everyday lives. Worldwide in 2010, according to the International Telecommunications Union, 640 million homes, housing 1.4 billion people, had at least one computer in the house; 525 million of those households and 900 million people were connected to the Internet. In the United States, about 90 million American households (80 percent of the U.S. total) had PCs and Internet access, and nearly half of those had two or more computers at home; 70 million had a game platform like Wii, PlayStation, or Xbox; 45 million households shared some 96 million smartphones; 7 million had tablet computers. Sixty percent of households had three Internet-enabled devices; a quarter had five.

Over the course of a typical day, you send and receive an average of 110 messages. You check your phone thirty-four times, visit Facebook five times, spend at least half an hour liking things and messaging friends. Like most people, your smartphone is more *smart* than *phone:* for every hour you spend talking to someone, you spend five hours surfing the Web, checking e-mail, texting, tweeting, and social networking. Nielsen and the Pew Research Center have found that Americans spend an average of 60 hours a month online, or 720 hours a year. That's the equivalent of 90 eight-hour days per year. Twenty of those days are spent in social networking sites, 38 viewing content on news sites, YouTube, blogs, and so on, and 32 doing e-mail. If maintaining your online life feels like a job, maybe that's because it is.

The increase in the number of digital devices we own and the amount of time we spend with them doesn't mark just a quantitative shift. It's a qualitative one too. Digital technologies and services are entwined with our everyday lives, whether we like it or not. As one Silicon Valley engineer put it, "Computers used to be part of my daily life. Now they're part of my daily *minute.*" A veteran of Google and Facebook, even she feels the change; like many of us, she's aware of information technology playing a larger role in the casual and necessary things we do to maintain our homes and family and social lives. People who spend all day with computers used to be called hackers. Today, that's all of us.

Digital life can be great, but it also has a price. Keeping up with everything that everyone's sharing can become overwhelming—not just the sheer volume of material, but also the obligation to stay on top of it. These are your friends (or "friends"), and if you don't keep checking in on what they share, you might miss something. The little buzz from a new text message or e-mail is nice, but it's also disappointing when you hit Refresh and nothing's there.

Sometimes the problem feels bigger. Having to stay focused when everyone wants your attention and the world—as well as your friends—throws a constant stream of distractions at you is *hard.* It's easy to get sidetracked at work by one thing, then another, and then have real trouble finishing the task you started. Recent surveys and field studies have found that a majority of workers have only three to fifteen minutes of uninterrupted working time in a day, and they spend at least an hour a day—five full weeks a year—dealing with distractions and then getting back on task. Each little thing you respond to feels urgent and gives you that sense of being busy, although you have the sneaking suspicion that all the interference and overlaps make you less productive. But when everyone looks perpetually busy, being overloaded is a badge of honor; working too hard is the new normal. Multitasking makes you feel like you're working even when it's counterproductive.

Organizations pay a price for their employees' chronic distraction. In a 1996 global survey of managers, two-thirds thought that constant distraction and information overload affected their quality of life. More recent studies estimated that in 2010, information overload cost U.S. businesses about 28 billion hours of wasted time and $1 trillion, and this was a year when the nation's gross domestic product was $14.6 trillion. The average worker spends half an hour a day troubleshooting devices or dealing with network problems. Over the course of a year, that's fifteen workdays lost to computer problems.

The constant buzz, the need to keep up with the never-ending rush of information, and the efforts to divide and spread one's time and attention ever thinner are starting to take their toll. It's getting harder to concentrate when you really need to. You reach the bottom of a page and can't always remember what you just read. Not only do you have trouble getting back to that task you started an hour ago, but you have to struggle to remember what that task was. You forget things on your mental shopping list. At home, sometimes you head to a room to do something and forget what you meant to do by the time you get there.

Now let's imagine a different Monday.

Monday morning. You reach over to the nightstand, grab your smartphone, and switch off the alarm. You don't check your mail or the online news just yet. After a few months of evaluating your mood when you check your mail first thing, you know you'll have a better day if you wait. Besides, you want just a little more time offline. Late Saturday night, after setting up your coffeemaker, you put your phone on silent and stowed the laptop and tablet in a desk drawer. Six days of the week you're connected; now you and some friends spend Sundays doing intensely analog things. Sometimes you head for hiking trails or cook; a couple of them have rediscovered knitting and painting. This Sunday was taken up by baking and reading. After a few hours at the market and some measuring and mixing, you had enough coffee cake to get you through the rest of the eight-hundred-page novel penned by the latest writer to burst from Brooklyn's literary scene.

When you do check your mail on your phone, you open the program, then put the phone screen-down on the table while you get coffee. It's a little act of resistance: *I'll get to you,* you're saying, *when I choose.* There's not much in your in-box, even after thirty-six hours away: you've turned off every notification, unsubscribed from all but the most useful newsletters, and have an aggressive set of filters that shunt nonessential mail out of your in-box before you see it.

At work, you need to be heads down, regardless of your colleagues' immediate needs. Yes, it's important to be responsive, but a frantic request isn't the same as a high-priority one, and you have work to do. So you switch off your phone and fire up a program that blocks your Internet access. For two hours, you've got no external distractions and no opportunity for self-distraction: e-mail, Facebook, Pinterest, Amazon, your colleagues—they all have to wait. If coworkers need something, they know where you are, but by making people invest a little effort to get your attention, you filter out the ones who want your time but don't really need it.

Now you have a single task, which you think of as kind of like a game: Produce this many words, or write this much code, or get through this many accounts. After a while, your mind settles into a groove. You feel a bit like a jazz drummer: totally engaged but on beat, not a movement wasted.

After two hours, you switch everything back on. It's amazing how much you can get done when you focus on one goal. It often still involves multitasking, but it's the sort of multitasking that converges on a single point, not the kind that pulls you in different directions.

In the evening, you give yourself half an hour to see what your friends are doing on Facebook and Twitter. Occasionally, you pare down your list of friends. Your timeline is less cluttered, because you're more careful about whom you give your attention to. In real life, your circle of friends contracts and expands, and the amount of time you can devote to people is constantly changing. You write fewer messages and check in less frequently, and you try to post things that are well composed and thoughtful. Your aim isn't to get noticed or to accumulate lots of followers. Being online is about connecting meaningfully with people, about conserving your attention and respecting your friends' minds, not about killing time or engaging with media for its own sake. More generally, you try to use information technologies as mindfully as possible. You observe what you do, see how different practices affect your productivity and mood, then take up better practices and discard obsolete ones. But when things go well, you can turn off that mental camera, feel a device go from a tool to an extension of yourself, and become completely absorbed in the moment.

Relating to and using technologies this way—practicing contemplative computing, in other words—requires understanding and applying four principles.

The first is ***our relationships with information technologies are incredibly deep and express unique human capacities.*** It sometimes seems that technology threatens to reduce us all to scary, soulless man-machine cyborgs like the Borg and the Terminator. But as Andy Clark, a philosopher and cognitive scientist at Edinburgh University, argues, we're really "natural-born cyborgs," forever seeking to extend our bodies and cognitive abilities through technology. In fact, it's best not to see the mind as something confined to the brain or even the body; it's useful to think of oneself as having an "extended mind" (to use Clark and David Chalmers's term) made up of overlapping parts that link brain, senses, body, and objects. Today's information technologies, I contend, cause us pain not because they're supplanting our normal cognitive abilities, which have always been flexible and mobile, but because they are often poorly designed and thoughtlessly used; they're like limbs that we can't bring under control.

The second big idea is ***the world has become a more distracting place—and there are solutions for bringing the extended mind back under control.*** Contemplative spaces are disappearing as quickly as tropical forests, work and life are becoming more frenetic, and modern technologies present challenges to one's ability to concentrate that may be unique. But humans have always had to deal with distraction and lack of focus—and for thousands of years, they have been cultivating techniques that effectively address them. In Asia, Buddhist and Tantric meditation, Japanese Zen, and Korean Son and yoga have all evolved to tame the distractible, chattering, undisciplined monkey mind. Neuroscientists, psychologists, and therapists have all observed that meditation practices can have a powerful effect on the brain; they can sharpen physical abilities and help deal with a host of psychological problems. Contemplative practices offer more than just a way to control the monkey mind or curb compulsive multitasking. They can also be adapted to allow you to regain control of your extended mind.

The third big idea is ***it's necessary to be contemplative about technology.*** You have to look closely at how you interact with information technologies and how you think about those interactions in order to understand how your extended mind develops and works. Our interactions with information technologies—with the outer reaches of our extended minds—are shaped by a variety of factors: the designs of devices and interfaces, the ways and contexts in which we use devices, and our mental models about the interactions and ourselves. Those models often carry unexamined assumptions about how information technologies work and how we work that are detrimental to us.

The fourth big idea is ***you can redesign your extended mind.*** Understanding the extended mind, having a better grasp of how to choose and use technology, and being familiar with contemplative practices let you find ways to be calmer and more purposeful when using information technology. It helps you be more powerful in exercising your extended mind and more deliberate in strengthening it. By understanding how all these pieces fit together, you can be contemplative through technology—and, in the process, regain your ability to deal with challenges, think deeply, and be creative.

Contemplative computing isn't just a philosophical argument. It's theory *and* practice. It's a thousand little methods, mindful habits informed by the four principles. Guidelines for checking e-mail in non-distracting ways. Rules for using Twitter and Facebook that encourage thoughtfulness and kindness. Ways of holding—literally *holding*—a smartphone so it commands less of your attention. Techniques for observing and experimenting with your technology practices. Methods for restoring your capacity to focus.

Information technologies are so pervasive, so much a part of work and home, so thoroughly embedded in modern life, it can be hard to know where to push back first. A good choice is to begin where many contemplative practices start. With breathing.

ONE

BREATHE

Before you read any farther, get your smartphone or iPad or laptop and check your e-mail. This is probably something you do multiple times a day; for many of us, it's almost a reflex, and we hardly think about it. Productivity experts recommend checking work e-mail only a certain number of times a day, but lots of people hit the new-mail notifier on the computer's menu bar or punch the Get Mail button in the e-mail program every few minutes. It's an automatic, nervous habit, like glancing at one's watch. Computers automate the practice and check for us several times an hour. If you have alerts running on multiple devices and several e-mail accounts, that can translate into hundreds of interactions with your in-boxes every day.

So go check your mail. But this time, as you do, don't think about the messages that might be waiting in your in-box or about how you really should have answered those messages from last week. Try not to let your thoughts wander. Instead, pay attention to yourself. Try to observe what you do. Watch how the computer reacts to you and how you react to it.

In particular, notice how you breathe. Did you hold your breath? Chances are, you did, and that small unconscious habit is a window into a big world of issues. It shows how a disembodied transfer of information that we think has nothing to do with the physical world actually does have a bodily, physical dimension. It illustrates how we don't use information technologies in the way we use bicycle pumps or elevators or salad tongs; the technologies turn into extensions of our minds and memories. They become entangled with us.

Linda Stone is a technology consultant, writer, and former Apple and Microsoft executive, the sort of person who can coin a phrase like *continuous partial attention,* which describes the way a person divides his focus among multiple devices, never giving any single one his complete attention. In 2008, she noticed herself holding her breath while checking her e-mail. After observing people at cafés and conferences, asking friends, and doing some informal surveys, she found that lots of people held their breath when they checked their e-mail.

Stone called the phenomenon e-mail apnea. The term is a play on sleep apnea, a breathing problem caused by either a physical obstruction in the airway that keeps air from reaching the lungs or a failure of the brain to signal the lungs to breathe. People with sleep apnea can stop breathing hundreds of times a night, sometimes for up to a minute. It's not usually fatal, but it can contribute to fatigue and impaired cognition, and even to physical problems like obesity and heart disease.

E-mail apnea is probably more pervasive than sleep apnea. Somewhere between 100 million and 350 million people worldwide have sleep apnea; in the United States, it's estimated to be as common as heart disease, clinical depression, or alcoholism. But roughly two billion people worldwide, nearly a third of the Earth's population, use computers. Roughly two billion people have broadband Internet access. More than twice that number have mobile phones.

It's not a stretch to assume that e-mail apnea, like sleep apnea, isn't very good for us. Stone speculates that holding one's breath while checking mail is triggered by the fight-or-flight reflex. It reflects the anxiety many of us feel as we check for new messages in our in-boxes, not knowing what new fires we'll have to put out or what problems we'll have to solve. We see variants of it in other electronic interactions: when you're waiting for a critical text message, for example, or when you unexpectedly have to update the printer driver in order to print out the document you really need for a meeting that's starting in a few minutes.

E-mail apnea is the kind of chronic condition that can make a person's life a little more unpleasant, and make a person more unpleasant to others. Those six billion devices making all of us a bit more anxious do connect us to one another, after all. But we're barely aware of the problem.

E-mail apnea shines a light on an important but usually unrecognized dimension in our relationships with information technology: the degree to which our minds, bodies, and technologies can become entangled. Researchers used to believe that the mind and consciousness emerge out of the brain's cognitive functions. But as they've gotten to know more about how the brain works and how the mind reacts to new technologies, some philosophers and cognitive scientists have begun to argue that the boundaries between mind and body, and even the boundaries between the mind, the body, and tools and surroundings, are pretty fuzzy. They argue that it's wrong to think of the mind as being contained by the brain. Rather, they propose a model of an extended mind, consisting of brain, body, devices, and even social networks. The extended-mind thesis argues that we need to understand cognition, or thinking, as something that can happen anywhere in this system; a person might internalize some cognitive functions in memorized rules or in his subconscious, outsource others to technologies, or use a combination of memory and device to get things done. Even something as apparently simple as reading turns out to be a vastly complicated ballet of unconscious processing and conscious action that's coordinated across body, book, eyes, and hands.

Homo sapiens has a very long history of entanglement; interactions with technologies change the way our bodies work and the way our minds work. Entanglement allows us to extend our physical and cognitive abilities; do things that we could not do with our bodies alone; accomplish tasks more efficiently, easily, or quickly; and achieve the kind of mastery that lets us lose ourselves in our work. It stretches the body schema, the unconscious mental map of where one's body ends and the world starts. This is why a common statement like "my iPhone feels like part of my brain" actually expresses some deep truths.

The term *entanglement* combines several phenomena that scientists and philosophers have studied separately. I prefer the term *entanglement* over other options for a couple of reasons. The term *extended mind,* coined by philosophers Andy Clark and David Chalmers, sounds a little too positive. Extending one's cognitive ability or memory sounds like an unambiguously good thing; we need a term that acknowledges that some intimate engagements with technologies feel less like extensions and more like constraints. We also need to recognize that even the most positive extensions come at a price; nothing in our relationships with information technologies is completely positive or negative.

Entanglement also suggests a degree of complexity and inevitability. We naturally distribute our cognitive capabilities across our brains and an array of devices, and we all use technologies to extend our physical capabilities; it's something we've been doing unconsciously almost since birth. We're stuck with our devices. But you *can* choose whether you're tangled *in* your devices, like a fly in a spider's web, or entangled *with* them, like a strand in a rope. The second creates something that's stronger than its individual parts. You know what happens to the fly.

The concept of entanglement might sound like a transhumanist fantasy, the sort of thing that leads to dreams of uploading human consciousness into computers. Certainly there are lots of people who would welcome a disappearance of the boundary between humans and machines; futurist and inventor Ray Kurzweil, for example, envisions a future in which robots and artificial intelligence are as smart as humans, nanoscale robots are able to map every atom in the brain, and mankind has moved from having a single consciousness in each person's head to having all minds distributed among bodies, robots, and the Cloud. But people *already* talk about information technologies as if they were extensions of themselves. Users often describe their mobile devices as being parts of themselves; they also describe themselves as being "addicted" to the Internet.

The popularity of these metaphors is illustrated by a pair of studies conducted by researchers from the University of Maryland. In 2010 and 2011, they recruited college students in ten countries to stay offline and away from all media for twenty-four hours. After putting away their cell phones, many reported feeling like a part was missing. "I reached into my pocket at least thirty times to pull out a vibrating phone that wasn't there," one American student said. A student in China said, "I just like touching my cell phone with my hands, which made me feel full." One British student "actually craved having my phone, and routinely checked my pockets for it every five minutes," while another described it as "very strange not to have my phone constantly connected to my hand." Many reported feelings of withdrawal. "This day is simply composed by struggle and suffering!" one Chinese student wailed, while another said, "After twenty-two hours living without any media, I can say without exaggeration, I was almost freaking out." One American student reported that he "felt like a drug addict, tweaking for a taste of information," and another that "[I] needed my electronic 'fix'" and that going offline "literally felt like some sort of withdrawal." A British student simply admitted, "I am an addict. I don't need alcohol, cocaine or any other derailing form of social depravity.... Media is my drug; without it I was lost." Psychologists in the United States now argue whether Internet addiction (a term first used in the scientific literature in the late 1990s) should be recognized as a medical condition like alcoholism.

Thinking in terms of extended minds and entanglement helps clarify what's at stake when our relationships with information technologies go bad. As devices moved from tools we used at work or in class to things we lived with, they became more deeply integrated in our lives, and their potential to affect the shape and workings of our minds grew. When a device doesn't work well, it isn't just an inconvenience. You experience the malfunctioning device as a part of you and, at the same time, as something outside your control. It is like a limb that won't obey your commands. The problem with too many devices today is not that they are too engaging or addictive. The problem is that they are poorly designed.

Knowing what entanglement is and how it works is a big step toward using computers in a more contemplative way. We can't begin to have better relationships with our devices until we have a clear picture of what better *is*. Entanglement teaches us that we shouldn't worry about becoming too dependent on technologies. Throughout history, *Homo sapiens* has been inseparable from technology.

Our protohuman ancestors first used stones as tools about two and a half million years ago; the Acheulean hand ax, a sharpened, pointed, versatile tool that required substantial skill to make, was invented about 1.8 million years ago, and variations of it remained among our ancestors' most prized possessions for more than a *million* years. (I've held million-year-old hand axes that still have their edge. Imagine any present-day technology even *lasting* a million years, much less still being usable and useful.)

Humans have literally never lived in a world without tools, and tool use in humans evolved in concert with both biological and cognitive innovations. Our ancestors' brains—particularly the frontal lobes—expanded dramatically about the time they began to make and use tools. This neurological expansion helped increase our ancestors' capacity to form abstract ideas about how objects could be used; to remember those uses; and to teach them to others. The production of stone tools in flint-rich lands for use in hunting or fishing elsewhere also provides the first evidence of our ancestors' planning for their futures.

Our species' external features have also changed along with our use of tools. The development of bipedalism created an opportunity for our ancestors' hands to specialize in feeling and grasping rather than walking. This in turn made it possible for protohuman hands to become more tool-friendly; evolution selected for hands that had shorter fingers, and nails rather than claws. (Recent studies show that apes are unable to make hand axes and other stone tools because their wrists are too stiff and their fingers too short.) However, these evolutionary changes also made humans more dependent in some ways: they needed tools for hunting and fighting, and materials such as leather for protecting the skin from rough surfaces.

For the past couple hundred thousand years, humans have eaten more meat than gorillas or chimpanzees, but our species hasn't developed the sharp teeth or fearsome speed of other predators. In fact, although meat has come to occupy a larger part of our species' diet, our teeth and jaws have become *weaker.* Why? Teeth haven't evolved to tear living flesh from moving prey. Evolution selected for teeth that allowed humans to more efficiently consume *cooked* meat; animals were killed using technologies like spears and traps, and then the meat was cooked over fires. We're also less furry than our primate cousins, and we walk and balance differently, allowing us to make use of two other ancient technologies: clothes and shoes.

The human body took shape in a world where arrows, spears, traps, and knives became the technological equivalent of killer jaws and massive haunches; humans could rely on fire to soften and sterilize food. Technologies have changed mankind's environment and diet, and human evolution reflects that.

The evidence of cognitive entanglement is limited, because archaeologists have been looking for it for a much shorter period, and physical evidence of cognitive changes is ephemeral. One form that we can trace over the past twelve thousand years, though, is the discovery, cultivation, and use of psychoactive drugs.

In their natural state, the plants coca and khat are low-level stimulants, and they probably helped humans for whom cooked food and clothing were novelties to ignore hunger and maintain alertness during long hunts. Drugs became stronger and more refined with the rise of civilization, trade, migration, and imperial expansion. Old World paleobotanical sources (microfossils and preserved seeds, for example) and artifacts like ceremonial bowls and burners suggest that by about 10,000 BCE, peoples in Asia were chewing betel nuts as a stimulant. Ephedra and cannabis were cultivated by Chinese farmers by 4000 BCE, while their European brethren took to growing opium. Two thousand years later, nicotine and alkali-based hallucinogens were in use in the Middle East and Europe. Cannabis spread along caravan routes from China to central Asia and India and thence to Africa, while opium moved in the opposite direction, into Asia and the Near East.

In the ancient Americas, "plants of the gods" and rituals for achieving altered states of consciousness were pervasive. Andean peoples prepared ritual drinks made from the hallucinogenic San Pedro cactus from 1300 BCE, and coca and caffeine-rich guayusa were grown and traded from at least 500 CE. Ayahuasca, the "vine of the souls," was popular in the Amazon. In and around the Caribbean, a snuff called *yopo* was widely used; in low doses it is a stimulant, while in high doses it becomes a hallucinogen. Central America, with its rich tropical forests, was a veritable pharmacopoeia. The Maya in what is now Guatemala were using sacred mushrooms like *teonanácatl* as early as 500 BCE, and shamans in the Mexican region of Oaxaca developed rituals featuring brews made from mushrooms, ololiuqui (a kind of nightshade), and peyote from 100 CE.

Other entanglements developed with the domestication of animals and the rise of agriculture, the growth of urban settlements, and the development of complex societies. The growth of long-distance trade and the appearance of far-flung political entities created a need for reliable communication and record-keeping, and that stimulated the development and use of writing in Asia, Mesoamerica, and the Near East. Writing supported social enterprises of unprecedented [illegible] also had a powerful effect on the human mind. As W[illegible]rably put it, "Writing is a technology that restructures thought."

Reading knits together regions of the brain that evolved for different purposes but connected around the challenge of recognizing and deciphering texts.

Writing also externalizes ideas, making it possible for people to abstract and analyze concepts in ways that are very difficult in cultures that do not have writing. The flourishing of Greek philosophy and science, for example, was preceded by the spread of literacy through cities on the Greek mainland and in Greece's colonies in what is today Turkey. Literacy supported the development of longer, more elaborate forms of argument built on a wider range of sources. Writing made it possible for a person to take a mental step back and examine how authors made their arguments, to analyze the rhetoric and logic they used. From this point on, even spoken language bore the imprint of cognitive attitudes supported by writing.

Ancient civilizations are where we see some of the first evidence of deep ties connecting humans and technologies. Beyond an individual's awareness of a device's value, there was a sense that the possessor knew that ownership and skilled use was transformative—in other words, a conscious, self-aware variety of entanglement. Some of the best evidence comes from the Mycenaeans, a Mediterranean civilization that flourished between about 1400 and 1100 BCE; their burial practices suggest that swords came to be treated as extensions of their users. A Mycenaean warrior didn't merely *use* a sword, says Lambros Malafouris, a tutor at Oxford University and one of the leaders in the field of cognitive archaeology; he became a "human/nonhuman hybrid" of person and weapon. Spears and, earlier, bows and arrows could be made by hunters or warriors themselves, but good swords had to be made by skilled metalworkers; they were often ornately decorated, and they were punishingly expensive. (Little wonder that while swords were used in very different ancient civilizations, ranging from the Greeks to the Japanese, all cultures shared the idea that swords possessed their own lives and spirits.) Mycenaeans treated swords with great care and buried them with their owners, suggesting that sword and warrior were seen to share a unique, profound bond, a deeper and more symbiotic relationship than a hunter had to ax or bow.

So entanglement is nothing new or revolutionary. It's what makes us human. Our sense of ourselves, our bodies, and our minds are all shaped by it. Indeed, our evolutionary success—surviving in a world of much larger predators and outwitting our Neanderthal and Cro-Magnon cousins, which allowed our species to completely spread around the globe some forty thousand years ago—depended on it. And entanglement is no less important today.

Let's consider a simple example of the physical, bodily dimension of entanglement: the effect of technologies on the body schema. (I should say *principally* physical entanglement, because there's no clear boundary between physical and cognitive entanglement; all entanglements that change the body also affect the brain and mind, and entanglements aimed at changing cognitive abilities often have a bodily component as well.) A body schema is the mind's model of the body: it tells you, for instance, how far your limbs reach, where you are in space, and how much volume you take up. Body schemata are important because they help us successfully operate in a complex world. Grasping a cup requires knowing how long your arm is and how wide you can spread your fingers, while walking down the stairs requires awareness of how far you can extend your leg without losing your balance.

Body schemata are flexible. Idoya's brain moved its robot-control functions away from its tool-using section and into its body section; when that happened, Idoya came to treat the robotic arm as indistinguishable from her other limbs, and her body schema expanded to include the robot arm. Even without something as dramatically high-tech as brain implants, people can become so practiced at using tools, the tools feel like extensions of themselves.

Consider the case of the blind man and the cane (a favorite among philosophers). A blind man is aware of his cane when he sits and holds it in his hands; he can measure its length, its weight, how flexible it is, and so on. Once he stands and uses it to move about, his awareness of the cane itself vanishes. His mind focuses on the information the cane's interaction with the space in front of him provides, and he feels as if he can sense objects several feet away. He treats it as an extension of his hand and "feels" the road ahead of him as if he were truly touching it. These remappings can happen rather quickly. For monkeys, the use of a tool as simple as a rake or a mechanical grabber (essentially a hand at the end of a long stick) to reach food alters the monkey's body schema within a few minutes.

The need for accurate body schemata and the requirement of flexibility create an opening that technologies exploit to make themselves part of the body. Given how much we have depended, individually and as a species, on successful tool use for survival, it makes sense that we would develop the ability to incorporate technologies into our body schemata. It's an efficient way to learn how to use technologies fluently—so fluently that you no longer have to think consciously about them and can focus only on the information they give you about the world or on their effects on the world.

Other tech expansions of the body schema are less useful. One example is phantom cell-phone vibration, the feeling that one's cell phone or pager is buzzing when it's not. A survey of medical workers at a Boston-area hospital revealed that two-thirds of respondents had experienced phantom cell-phone vibrations (or "ringxiety," as psychologist David Laramie calls it). People who regularly carry their cell phones in shirt or pants pockets, near the nerve-rich areas of the breast or upper thigh, are most susceptible.

What causes this? Scientists think that as a person becomes used to having the cell phone buzzing against his skin, his body begins to misinterpret a brush of clothing, a bump against furniture, or even a minor muscle spasm as a cell phone's vibration. It seems that for regular users, cell phones "enter into the neuromatrix of the body," as neuropsychologist William Barr puts it. "If you use your cell phone a lot, it becomes part of you." It's even more likely to become part of you if you can't afford to miss calls. In the Boston-hospital study, students and house staff checked their pagers and phones constantly and were much more likely than senior faculty to feel phantom cell-phone vibrations. Medical students' nervous systems had internalized the fact that the cost of a false positive (being aware of what turns out to be a nonexistent call) was far lower than the cost of a false negative (being unaware of a real call) because, one senior physician explained, "all hell breaks loose" if the students don't answer their phones.

Students who went offline for a day in 2010 as part of a University of Maryland experiment widely reported having ringxiety. "I definitely felt some psychological effects, such as hearing my cell phone ring even though it was off," one student said. "The experience of 'phantom ringing' was a little disturbing," another admitted. "It really enforced my dependence on my phone." Surveys from the late 2000s of cell-phone users in places as different as Iraq and California found that about 70 percent of people felt phantom cell-phone vibrations. This suggests that something like three *billion* people worldwide may feel their cell phones going off when they're not. Those numbers may be rising. In a 2012 study of American undergraduates, 89 percent of respondents said they felt phantom vibrations every couple of weeks.

It's not unusual for a human to work so intimately and effortlessly with an instrument or machine—a calligraphic brush or a motorcycle or a sword—that the device ceases to feel like something being used and comes to be experienced as an extension of oneself, another sense through which one interacts with the world.

Entanglement often happens without our knowledge or awareness, but we can become very conscious of our abilities, our sense of space or physical boundaries, and even ourselves, being stretched or strengthened in the course of using technologies.

This happens with musical instruments: a clumsy awareness of strings and valves and chord positions eventually gives way to a sense that the instrument "effectively becomes a natural extension of yourself," as one jazz musician put it. Another musician described the experience of mastering an instrument as acquiring "a new, beautiful, portable voice." We become more conscious of entanglement when we need formal training or deliberate practice to make it happen. Pilot and military historian Tony Kern wrote that, likewise, aviators need "knowledge, understanding, and trust of the airplanes" and a "genuine desire to make the machine an extension of yourself—a real attempt to bond human and machine into a single functional unit." With many technologies, practice delivers familiarity and basic competence that later becomes a foundation for deeper skill. Over time, your awareness of the device itself fades, just as your awareness of your enhanced abilities increases.

We also become more aware of entanglement when novel technologies let us do things that unaugmented human bodies cannot.

Nineteenth-century accounts of a new machine, the bicycle, highlight this. The bicycle "is worse than useless until animated by the guiding intelligence of which it becomes the servant and a part," an anonymous author wrote in 1869. It "increases your sense of personal volition the instant you are on its back," the writer continued. "It is not so much an instrument you use, as an auxiliary you employ. It becomes part of yourself." Thirty years later, another wrote that "the machine is an extension of yourself. On any other vehicle you are freight. Here you are moving by your own will and strength." A motorcyclist from that same period wrote, "As you fly along there are no longer two factors—yourself and the machine—but rather a single entity, a combination of flesh and blood and metal, in perfectly harmonious relation making a laughing stock of space." The bicycle may be the first machine that users described in such intimate terms; if so, its invention is an unheralded milestone in cyborg history.

You become acutely aware of this extension when your human/nonhuman hybrid is different but still feels like "you"—not exactly the same person as you were before but a version of yourself with the power to express things you wouldn't normally be able to. Georgia O'Keeffe's widely cited declaration "I could say things with color and shapes that I couldn't say any other way" is one that many painters immediately understand. Musicians and artists speak of being able to express ideas in sound or on canvas that they could not express in words. This sense of transformation isn't confined to artists; drivers and fighter pilots also write of being "part of a beautiful machine" that lets them "transcend the limitation of the human body" and move with a speed and power that they never could on their own.

At its most intense, entanglement dissolves the awareness of any difference between person and object: you work with it so perfectly, it becomes impossible to tell where you end and the device begins. This state has been described for centuries by masters of Zen arts. As Eugen

Herrigel related of his experience with Zen archery, "In the end the pupil no longer knows which of the two—mind or hand—was responsible for the work." Near the end of several years' studying archery in Japan, he describes how "bow, arrow, goal and ego, all melt into one another, so that I can no longer separate them." Avid turn-of-the-century bicyclists and motorcyclists described riding in similar language. The 1909 motorcycle rider gushed, "Where the conditions are ideal the state of your mind is indescribable.... [You become] part of your machine, while the machine—well, it's part of yourself." A 1904 bicyclist wrote, "The great charm of it all is indescribable. You and your wheel are one.... You feel that your joy must be akin to that of the eagle's flight, the very poetry of motion. You forget all about destination, rate of speed, extension of muscle; and you swing along exulting in the flight itself." To the modern mind, this description seems strikingly familiar, and its account of intense concentration, loss of self, and distortion of time perfectly describe Hungarian psychologist Mihaly Csikszentmihalyi's concept of flow. The ability to merge one's awareness and body schema with a device, be it a hand ax, violin, or F-15, is one that the body richly rewards.

That sense of merging with a machine doesn't come only when a person is sitting in a jet fighter cockpit. Programmer Ellen Ullman describes the feeling of being "close to the machine" when hardware, the programmer's mind, and code all fall into a beautiful, energizing alignment. It starts with the first glimpse of a solution to a difficult problem; at that instant, "Human and machine seem attuned to a cut-diamond-like state of grace," she writes. "Once in my life I tried methamphetamine: that speed high is the only state that approximates the feel of a project at its inception. Yes, I understand. Yes, it can be done. Yes, how straightforward. Oh yes. I *see*." Programmers can't just solve problems in their heads; the solutions may seem clear in principle, but writing good code is difficult. Seeing an elegant solution isn't the same thing as making a working product. To make the jump from an idea to a code that runs, "the programmer has no choice but to retreat into some private interior space," Ullman says, a place "where things can be accomplished."

The canonical picture of the completely absorbed programmer always has him or her at a keyboard, typing code. Connecting with the keyboard sparks ideas that some programmers can't have anywhere else, and there's tacit or technical knowledge that they can express only through the keys. When mathematicians work at a blackboard to solve highly elaborate theorems, they're not using the board just because it's convenient. The blackboard profoundly extends their short-term memory, helps them visualize the problem-solving process, and makes any errors more obvious; the problem-solving doesn't happen just in the mathematician's mind or just at the blackboard but in a cognitive system formed by both. Likewise, I suspect that the knowledge that programmers draw upon when they're close to the machine doesn't flow only from their brains; rather, it's distributed among brains, hands, and keyboards.

I suspect this because I have my own example of distributed cognition: I spell with my hands. I learned to touch-type as a child, and after years of lessons and decades of practice, I can close my eyes and touch-type upwards of seventy words a minute. Just as fluent readers are able to recognize entire words rather than letters, I type based on a tactile sense of how a properly spelled word feels. I can tell how my fingers should move and how my hands are supposed to angle and tilt as I roll through a set of letters. I can instantly feel if that pattern has been interrupted by a misplaced keystroke. I can't always identify the typo by feel—I have to open my eyes and look at the screen to do that—but I can almost always tell that it's there.

Because of this intimate familiarity with the keyboard, when my children ask me how to spell a really long word, I don't visualize an imaginary dictionary and read each letter. I observe the pattern my fingers make on an imaginary keyboard and then read out the letters. There are complicated words or names that I can't reliably write with a pen or on a touchscreen's virtual keyboard (where my decades-developed cognitive/muscle memory is rendered useless) but that I can type accurately on a full-size keyboard.

Like any kind of cognitive offloading, coding memories as manual activities has tradeoffs. People who navigate using visual landmarks rather than street names have a hard time describing routes to others. The obvious downside is that the talent doesn't generalize. If the keys for punctuation are in different places on the keyboard, it's enough to slow me down substantially, as I'm reminded whenever I have to use a British keyboard, which has nonletters in unfamiliar places. On a French or Japanese keyboard, with more letters rearranged, I become hopeless. The tiny keyboards on a smartphone are somewhere in between: I can slowly reproduce the patterns of words, but I can't rely on the full muscle memory—the tilt of my wrist or extension of my fingers—to know instantly if I've made a mistake. The advantage to my spelling with my hands is that it lets me write very quickly; at my fastest, I can type as quickly as I think.

I'm not alone in calling on motor memory when I need to remember something. Many people recall phone numbers and passcodes by imagining the pattern the fingers make on the keypad rather than by memorizing strings of numbers. This was hard to do in the days of rotary-dial phones, but with keypads, it becomes simple. We used to let our fingers do the walking; now, we let them do the remembering. One of the more striking examples of intellectual knowledge being encoded in the hands is the case of a typesetter at Oxford University Press who caught an error in a Greek book he was typesetting. He didn't speak or read Greek, he explained to the book's editors, but he had been setting books in the language for decades, pulling type from trays and arranging the letters in print frames, and he knew he had never reached for that combination of letters before. It just felt wrong. And it turned out he was right.

Offloading isn't used only for cognitive-manual activities; we constantly outsource memories to technologies, our surroundings, and other people.

For years I've carried around a pocket-size Moleskine notebook. I started this practice more than twenty years ago, when I was in graduate school. I had been an irregular diarist before then, but a few months into my dissertation, I had regular bouts of déjà vu; I'd spend an hour in the library tracking something down, then have the nagging feeling that I had worked on this very question a couple weeks before. In order to keep better track of my work, I started using the historian's equivalent of a laboratory notebook.

My small notebook is now a stream-of-consciousness transcript of my days. For example, two facing pages have a note about the Latin etymology of *contemplation,* a reference to a prototype location-based to-do-list system, the names of people I need to interview, a grocery list, directions to a restaurant in Cambridge, notes from a lunch with cognitive archaeologist Colin Renfrew at said restaurant, a William Blake quote copied from a show at the Tate, and, for some reason, my son's Social Security number.

The notebook also has a few other traces and tools: ticket stubs taped onto the pages, business cards I collected at the last conference I went to, stamps and more business cards in the pocket, Post-its on the inside front cover. A map of the London Underground is taped to the inside back cover, both to help me navigate one of the world's biggest subway systems when I'm in England and for nostalgia when I'm in the States.

The notebook itself has a slight curve from months in my back pocket. It'll last a couple more months before either I reach the last page or it rebels against its mistreatment and self-destructs. It's clearly a part of my daily life, but is a notebook just a tool? Philosophers Andy Clark and David Chalmers would argue that no, it's part of one's mind. It doesn't matter whether a shopping list is memorized or written down in a notebook that you carry all the time. "There is nothing sacred about skull and skin," as they put it (in a slightly creepy turn of phrase), when it comes to cognition and memory. What matters is whether the information—or the process—is easily available and trustworthy. They considered the case of Otto, an elderly man with Alzheimer's disease who wrote down lots of facts in a notebook that he always carried with him. Because the notebook offered Otto a "high degree of trust, reliance, and accessibility," Otto and the notebook were entangled.

Whether we realize it or not, we make decisions all the time about how we remember things. Often we use combinations of means to remember written or spoken texts. Roman orators developed amazing, elaborate tools for memorizing long speeches that involved constructing detailed visual reminders of critical points and then arranging them in an imaginary space. To recite a speech, you would walk through the space in your mind, letting each imagined object remind you of your lines. This may sound harder than just memorizing the speech, but it had the virtue of allowing the orator more flexibility in delivery than simple memorization would. Actors may have to memorize thousands of lines of dialogue for a play, and they often use a combination of external cues—their location onstage, how they're standing, what they're doing—to help them remember their lines and become familiar enough with their costars' lines to interact onstage convincingly.

So it was really no surprise when Columbia University professor Betsy Sparrow found that students used different memory strategies to learn material they would have to know for a test depending on whether they would have access to the Internet when they were tested on it. Students who wouldn't be able to go online remembered the information itself. By contrast, students who were told they could go online during the test remembered where and how to look for the facts rather than the facts themselves. As Sparrow and her colleagues noted, the Internet is becoming part of our transactive memory (transactive memory is a combination of memory stores individuals hold directly and memory stores individuals can access from elsewhere).

Do Sparrow's findings mean that students are getting dumber? According to some responding to her findings, the answer is yes. The *Guardian* described her work under the headline "Poor Memory? Blame Google." Another Web site ran with "Google Is Gradually Killing Our Memory Power" and suggested that depending on search engines was a bad idea "if you want to stay mentally agile."

But this argument is itself stupid. First of all, Sparrow's students were tested on trivia questions and factual statements. They weren't offloading the kinds of memories that we think of as defining who we are—the Proustian rushes of memory that are stimulated by an old picture or smell, the irreplaceable memories of holding your first child for the first time, or the recollection of dramatic turns, like Rick reading Elsa's farewell in the rain as his train leaves Paris. Further, transactive memory doesn't cover information itself; it relates to knowledge of how to *find* information. We use transactive memory constantly, and it can be incredibly efficient. Indeed, we live in a world supersaturated with triggers for transactive memory; they're called signs. We put them on the sides of buildings, on street corners, packaging, labels, and a hundred other places. We embed information in the world all the time.

Much of that embedded information is very place-specific. In my freezer, there are always a few family-favorite dishes ready to be popped into the oven. Although I warm them up frequently, I don't remember the oven temperature for each one or how long each should be cooked. It's not that I'm incapable of remembering; I enjoy cooking, and I'm a decent if limited chef. No, I don't even try to remember that the orange chicken cooks at 400 degrees for twenty minutes while the quiche cooks at 375 degrees for eighteen minutes, because I know those instructions are always on the side of the box, and I never need that information unless I have the box in my hand.

We also constantly use other people as repositories of transactive memory. When you ask a coworker a question rather than looking for the answer on the corporate intranet, rely on your daughter to remember the name of the next book in that vampire series that she and all her friends are reading, or trust your spouse to remember your flight information because you know he (or she) is always paranoid about that sort of thing, you're using transactive memory. Additionally, we have places whose very structures and layouts are gigantic tools for spatially organizing information, and we have places that link layout to information flow. They're called libraries and offices, respectively.

The idea that technologies can become extensions of the mind may still be a little abstract. How, you might wonder, do devices make everyday cognitive activities easier?

Let's look at something you're doing right now: reading. Reading has the virtue of being both very familiar and very complex and multilayered. By deconstructing it, we can see more clearly how cognitive functions that we develop with years of practice, formal techniques that we consciously learn and apply, and the physical nature of the printed page and book all work together. Taken apart, reading turns out to be an extraordinary combination of conscious and unconscious activities and internalized and offloaded processes, all integrated so well that they create a completely seamless experience.

First, observe something really basic: You're reading letters.

You recognize each one, you associate letters with sounds (this is called phonemic awareness), and you know how these sounds build to make words.

But you don't consciously string letters and sounds together. After years of practice, you're good at automatically grouping letters into words, because there are specific parts of your brain—the temporoparietal and occipitotemporal subsections of the left hemisphere and the speech-focused left inferior frontal gyrus, also called Broca's area—that have devoted themselves to phonemic processing. Functional magnetic resonance imaging (fMRI) studies of the brain indicate that when a person is learning to read, the temporoparietal region, which concentrates on letter recognition, works hardest. As the more conceptually focused occipitotemporal becomes more engaged, reading becomes faster and more fluid. When we read silently or encounter new words, the inferior frontal gyrus gets more involved, since decoding new words often requires sounding them out.

You're aware of reading words and lines of text, but what you don't realize is that your eyes aren't moving evenly across letters and spaces; rather, they're focusing on groups of letters for a couple of tenths of a second, performing these saccadic jumps without your awareness. (Your visual system learned to move your eyes like this, and when you were quite young, your brain learned to take these individual frames and convert them into a smooth picture of your visual reality.)

So word recognition is fluid and automatic, but it isn't an ability you were born with; it's one you've acquired over years, and it's moved from the conscious to the unconscious part of your brain. In the parlance of the extended-mind thesis, word recognition has been outsourced as an autonomic function.

Another thing that makes perceiving and recognizing words easier is that there are spaces between them.

Haven't noticed those since you were a child? You should. Incredibly, there was a time when word spacing would have been considered unnecessary, at best, or even a concession to weak readers. For Roman orators, texts were meant to be read aloud, not scanned silently, and only semi-literates needed the crutch of word spacing to help them decipher good Latin. It's a challenge to find individual words in long strings of characters with no spaces; this is what makes games where you look for words in a grid of letters entertaining. During the Middle Ages, word spacing was adopted to help provincial converts with a shaky grasp of Latin read the Bible and to help scholars make their way through recent translations of Arabic scientific and philosophical texts. For novice readers, word spacing made a new language easier to comprehend; for experienced readers, it made reading much faster and largely eliminated the need to read aloud. Reading could now become a silent, contemplative activity, less like speech and more like thought.

Now, return your attention to the letters. They may have little curls, and the spacing between the letters may vary ever so slightly. The curls are serifs, and they're used to make letters easier to read (although, five hundred years after the serifs' introduction by Venetian printer Nicholas Jenson, professional typographers still argue over their utility and aesthetics). The spacing varies because different letters need different amounts of space around them to look good. With a few notable exceptions, the typefaces and fonts used in books and magazines are designed to be easy to read.

Words tend to be printed in black on white or slightly off-white paper. Notice anything else about the letters you're reading? Some of them, particularly the ones at the beginnings of sentences or the beginnings of names, are bigger than others. Mixed in with the letters are punctuation marks, like commas and semicolons, which give your inner reading voice cues about how to read a sentence—when to pause, when to provide a little emphasis (or when something is meant as an aside).

Now look around the page. Notice the white space separating the words from the edges of the page. These margins make it easier for your eyes to scan lines and keep your place, and they provide space for you to make notes or annotations. Many books will also have running heads, words at the top of the page that communicate some information—the book title, for example, or the name of the current chapter. Each page also has a unique number. A few at the beginning are in Roman numerals, and the rest are in what we commonly but mistakenly call Arabic numerals (they were actually invented by Indian mathematicians, but Westerners discovered them through Arab scientific texts).

If you leaf through the book, you'll see other pieces of organization. At the front is a table of contents, which shows you (thanks to the page numbers) where new chapters begin. At the back of the book is an index, which shows you (again, thanks to page numbers) where in the book different topics are discussed.

These features will all be familiar, and you're accustomed to seeing them in books. In this book, they've probably received very little attention until just now. These structural elements are what bibliophiles call *paratexts,* a term that also describes headings and subheads, picture captions, and footnotes. Most of these have been features of books for hundreds of years; word spacing and punctuation are medieval innovations, and modern typography began as a real art in the Renaissance, when publishers started to use typefaces and designs to produce books that appealed to nonmonastic and nonscholarly readers.

Young readers don't see many of these paratexts: *Pat the Bunny* doesn't have chapters. Paratexts are intended to support reading that's more complicated, sophisticated, and varied than the sort practiced in the nursery. Reading serious fiction requires stretching your imagination, placing yourself in the minds of others, and expanding your emotional and sympathetic faculties. In college philosophy classes, we learn that reading means identifying an author's argument, evaluating the evidence, and becoming aware of the writer's rhetorical tricks. (In this, we live in the shadow of Mortimer Adler, whose classic *How to Read a Book* described two kinds of advanced reading: analytical and syntopical.)

Professional reading is even more targeted and opportunistic. In graduate school, my classmates and I learned how to read just enough of a book to understand its main argument, see its place in the scholarly literature, and assess its importance. We learned to think about books in a way that would later structure our thinking about our own work: "reading" was narrowed to intensively analyzing a few pages and quickly browsing others, and sometimes it included reading the book's reviews or the author's earlier work. Lawyers learn to read just as purposefully. Judges and experienced lawyers read opinions far more quickly and efficiently than first-year students do, because they more effectively use structural signposts, footnotes, and keywords to get a feel for an author's reasoning and use of precedent; they can zero in on parts of a ruling that are novel or controversial and quickly assess the impact a decision may have.

These kinds of reading aren't just techniques for managing large volumes of content; they direct an academic's or professional's attention and help define what it means to be a scholar or lawyer. But all these complex cognitive activities—keeping track of an argument, appreciating an author's command of language, experiencing surprise, interpreting the meaning of a ruling—rest on automatic, foundational abilities we develop as children. As Maryanne Wolf notes, reading takes milliseconds, and it takes years—the neurological side of reading becomes blazingly fast, while the cultural and interpretive elements of reading develop far more slowly.

If letters, words, word spacing and punctuation, fonts and typography, and paratexts are so commonplace, why are they worth remarking on? Because, despite their near invisibility, they help you achieve remarkable feats of recognition, cognitive outsourcing, and understanding. Letters, word spacing, and punctuation help you read and decipher words quickly and fluently. Your eyes are moving in rapid bursts over groups of letters, your brain's visual processing center is creating a smooth experience out of jagged fragments of visual data, and a nearby section is recognizing words. Chapter headings, page numbers, and footnotes show you at a glance where you are in a book, signal what you should be paying attention to, and tell you what content is central and what is peripheral. But all of this is happening at an unconscious level. Consciously, you're reflecting on the meaning of sentences, keeping the past few lines in your short-term memory, thinking about how the paragraph is structured, how this section's argument is ordered, and what it means. Pieces of the argument—particular facts, maybe a turn of phrase—are starting to settle down into long-term memory. You may be underlining or taking notes, creating a record that you can go back and refer to or that may help you better digest the book's argument.

In short, as you read, you interact with layers of technologies, from letters to margins to chapters, drawing on skills you mobilize unconsciously and automatically. This helps you orient yourself in a sustained, elaborate argument; lets you filter what's really critical from what's interesting but peripheral; and supports your efforts to turn reading into meaning and memory.

Finally, the tools in the book itself aren't the only tools we use. We add our own tools to it. Lots of people underline in books or make notes in the margins. When I start a book that I need to read carefully, I put Post-its in the inside cover and keep a pen nearby, and I intensively underline, annotate, and make notes as I go. Treating reading as a martial art helps me engage more deeply with the book's argument, keep track of its twists and turns, understand the author's strategy (or deceptions), and figure out what I really think of the book.

Other manipulations are much more prosaic. When you put down an unfinished book, you often outsource the job of remembering where you stopped to a thing—a bookmark. The bookmark has no memory per se, but you know from its location that *this* is where you stopped reading; you don't have to remember the number of the last page you read (there are those page numbers again) or the name of the chapter you'll start when you pick up the book again. (Though if I'm reading a story with my son, he will usually remember the action in the previous chapter we read.) If it's a receipt from a bookstore, a train stub, or a concert ticket, the bookmark may carry its own set of associations. These practices don't all work toward the same ends; I take notes on a book because doing so helps me remember its argument, but I don't use bookmarks to help me memorize page numbers.

The miraculous complexity of this process becomes apparent only when it goes wrong. Some people find it harder to submerge word recognition into the realm of the automatic. Children who are dyslexic have trouble with the order of letters, and this impedes their ability to connect words on a page with words they speak and use every day. There appear to be neurological foundations for dyslexia; studies show that when dyslexics read, their temporoparietal and occipitotemporal subsections are less active than those of nondyslexics. But that does not mean those brain regions have to stay that way—those areas become more active in dyslexic children who are in special reading programs. My son is dyslexic, and when we first tested him, we found his verbal and reasoning skills were off the charts, while his formal reading ability was well below normal. But he's had years of tutoring, and now his left hemisphere is catching up to his inferior frontal gyrus, and he's approaching the same reading level as his peers. Neuroplasticity and stubbornness are wonderful things.

In adulthood, even the most able readers have moments when they become more aware of the basics of reading. When you encounter a long, unfamiliar word, for example, you might break the word into syllables and then sound it out. When we learn a new language, we're conscious again of the struggle to puzzle out sounds and associate words with meanings; we become aware again of how valuable the ability to instantly recognize words is, how much effort can go into that normally invisible activity.

In fact, a literate brain can't *not* read letters and try to form them into words, which can cause problems if you visit countries that speak unfamiliar languages but use alphabets similar to yours. As an English speaker, I was never so confused by signs as when I visited western Finland, where the signs are printed in Finnish and Swedish, neither of which I speak. Unlike German, which shares many roots with English, or Romance languages, which have contributed lots of loan words to English, Finnish is completely alien, and I could make no sense of it.

By contrast, when I visit Korea or Japan, busy streets full of neon characters leave my reading brain untroubled, because I can't read hangul or kanji. (The guilt center, however, kicks into high gear as I consider how disgusted my grandmother would be at my inability to access my own cultural heritage.)

Most of the time, though, what we all experience when we read is a seamless blend of complexity and cunning. I. A. Richards was a lot closer to the truth than he realized when he wrote that "a book is a machine to think with." Reading is a form of entanglement, the best everyday example of an effortless but mind-expanding merger with technologies. A book contains layers of cognitive engagement, content and paratexts meant to exist at the center and periphery of our attention, tools that invite us to engage or offload. Reading does not involve total attention to every element of a book; we focus our attention intensively on certain things, rely on devices to help us remember others, and completely transfer responsibility to other objects for yet other information. All these literary technologies are invisible for the same reason that the lenses of your eyeglasses are invisible: you cease to be aware of them because you see the world through them.

* * *

As reading shows, becoming so familiar with technologies that they become part of ourselves, being able to use them effortlessly, and feeling them extend our physical or cognitive or creative abilities can be intensely pleasurable. You can experience the same feeling driving or cycling in those moments when you feel like the machine is an extension of your body and you're connected through it with the road. You can have the same sensation playing a sport or game, those times when the racket or controller becomes a part of your hand and you react to new threats before you realize it, feeling challenged but in control. You can feel it when rock climbing or hiking, when every sense is absorbed by your surroundings and your body is stressed but you don't feel like you're going to crash; rather, you feel like you're going to crash through your old limits.

That state is what Mihaly Csikszentmihalyi calls flow. Flow has four major components, Csikszentmihalyi writes. "Concentration is so intense that there is no attention left over to think about anything irrelevant, or to worry about problems. Self-consciousness disappears, and the sense of time becomes distorted. An activity that produces such experiences is so gratifying that people are willing to do it for its own sake, with little concern for what they will get out of it, even when it is difficult, or dangerous."

You can reach flow doing almost anything. Csikszentmihalyi (it's pronounced "*Chick*-sent-me-*high*-ee") has been studying flow for decades, and he and his collaborators have interviewed or surveyed thousands of people around the world, people of various different ages and in a wide variety of occupations. "We find people who have to slice salmon all day for lox and bagels in Manhattan, who approach their work with the same sense of creative commitment as a sculptor or a scientist," he tells me over Skype from his office at Claremont University's Drucker School of Management, outside Los Angeles. When he speaks, Csikszentmihalyi sometimes closes his eyes to concentrate on his words. A huge stack of books stands behind him, competing for wall space with framed awards and book covers.

[illegible]ters reach flow? "They say, 'Each of these fish is different: I usually have five or six salmon each day, and when I pick each one up and drop them on the marble surface, I develop a three-D X-ray of how this fish is made inside.' Then they can cut it with minimum effort, making the tiniest and most ephemerally fine slice, and leaving the smallest amount of flesh to throw out." It becomes a kind of game: get the most lox with the fewest cuts and the least waste.

Situations in which there are challenges, clear rules, and immediate feedback are likely to support flow. This is one reason that games—be they board games, like chess, or video games—are so appealing: players can get into flow states quickly. With simpler video games, Csikszentmihalyi says, "You have aliens and you have to shoot them. That [illegible]ger finger and an ability to react quickly." Jobs where people can construct short-term goals for themselves—rotate these three sets of tires, write five pages, load this cargo in a way that balances the weight and keeps the ship stable—keep their attention long enough for them to enter a flow state. (Having the autonomy to define one's goals also helps create a sense of independence.) Indeed, finding those goals, defining them in ways that make them challenging but achievable, is itself a skill and a sign of real mastery.

Games and tasks that are easy to master may produce flow more quickly than difficult ones, but they can't hold people's attention for very long. An activity like painting or practicing medicine, by contrast, may take years to master but can be engaging and challenging for a lifetime. Guitar Hero is relatively easy to learn and can be lots of fun, but after a couple hundred hours, the challenge evaporates. With a real guitar, there are always new songs, new styles to master, new ways to express yourself, even after decades. In situations where "the challenge is high, you start with low skills, but if you develop them, you can get into flow," Csikszentmihalyi explains to me. "In chess, bridge, and other highly complex games, for example, it may take many, many years before you exhaust the challenge. But flow is rarer because it's difficult to achieve."

Csikszentmihalyi and his colleagues in the field of positive psychology—essentially, the science of happiness—have discovered that people are happiest when they are absorbed in difficult tasks, not when they're diverted by sybaritic pleasures. "The best moments in our lives, are not the passive, receptive, relaxing times," Csikszentmihalyi writes. They "occur when a person's body or mind is stretched to its limits in a voluntary effort to accomplish something difficult and worthwhile." Challenge, exhilaration, worthwhile and rewarding difficulties, and an intense awareness of them: these are what produce flow, and flow is the key to happiness. The intensity of flow experiences help people understand who they truly are. "When you pay attention to the point that it really reflects who you are, what you've done, what you want to do," Csikszentmihalyi tells me, "you fulfill your role in this world; you feel good about yourself and your work."

The ability to *pay attention,* to control the content of your consciousness, is critical to a good life. This explains why perpetual distraction is such a big problem. When you're constantly interrupted by external things—the phone, texts, people with "just one quick question," clients, children—by self-generated interruptions, or by your own efforts to multitask and juggle several tasks at once, the chronic distractions erode your sense of having control of your life. They don't just derail your train of thought. They make you lose yourself.

When it goes well, entanglement lets you use technologies skillfully, even effortlessly. At its best, entanglement gives you tremendous pleasure, extends your ability to imagine and create, and gives your life depth and meaning. This explains why bad entanglement is so painful, why distraction is so corrosive, and why it's so important to have technologies that help you focus, be mindful, and flow.

One of the things that supports Zen-like, effortless use of technologies is steady breathing. In his classic *Zen in the Art of Archery,* Eugen Herrigel described the critical role breathing played in Japanese archery. Archery is an embodiment of Zen, Herrigel argued, an "artless art" that can be practiced only with a clear, mirrorlike mind. Breathing properly turns out to be as important as handling the bow correctly. If we do not conquer e-mail apnea while using information technologies, good entanglement will remain elusive. Fortunately, scientists are starting to experiment with ways to encourage better breathing during computer use. To learn about them, I visit the Calming Technology Laboratory, a research group at Stanford University led by doctoral student Neema Moraveji. I find him on the first floor of Wallenberg Hall, a sandstone-clad building in the university's main quad.

Talking with Moraveji is the closest I've come to interacting with a character from the science-fiction show *Lost:* He has the appropriately exotic biography, personality, and good looks to be a passenger on Oceanic 815 and the technical skills to be part of the shadowy Dharma project. The son of Iranian immigrants who moved to the United States in 1979, Moraveji studied computer science at Carnegie Mellon and worked at Microsoft Research Asia before coming to Stanford. He spent years backpacking in Asia and Latin America and can explain the benefits of meditation in several languages. The morning we met, his Facebook page had pictures of him with his fashion designer fiancée at Burning Man, a weeklong art event in the Nevada desert that's a favorite of Bay Area avant-garde artists and technologists.

The Calming Technology Laboratory sounds lofty, but it turns out to be Moraveji's laptop, some Web sites, a roving band of fellow thinkers, and several prototypes. When I meet Moraveji, he's wearing one of his recent projects, a chest sensor and an Arduino electronic controller (a favorite with tinkerers, thanks to its low cost and flexibility), and he has it hooked up to his MacBook. Other people in the lab are developing systems that piggyback on SMS messaging, digital photography, even Facebook. This is a field where you can do interesting experiments and proof-of-concept prototypes on the cheap.

The Calming Technology Lab develops technology to reduce the impact of everyday low-level stressors (things that trigger stressful reactions). The researchers don't want to eliminate useful or desirable stressors; stage actors and ER doctors learn to work very well under pressure, and adrenaline junkies pay good money to leap out of airplanes or ski down moguls. Rather, the lab is targeting low-level chronic stress, the kind that's produced by everyday friction and frustration.

I ask how they define *calm.* "What *calm* means," Moraveji says, "is 'restful alertness.'" Calm, focus, and attention are all connected; the higher your chronic stress rate, "the more distracted you're likely to be, and the less you can focus deeply and be productive," Moraveji continues. "And those distractions are reflected in the breath."

A change in breathing, Moraveji realized when he first started this work, doesn't have to be involuntary. Our breathing is often unconscious, and, as e-mail apnea shows, it can be affected by tasks or environment. But unlike heart rate and blood pressure, which also rise and fall with stresses, breathing *can* be controlled if people focus on it. Moraveji himself spent years mastering breathing meditation. "The breath is where mind and body meet," he says. "It's a simple mechanism for modulating one's state." It also has the virtues of being easy to measure, monitor, and quantify—which make it a good candidate for digital intervention.

The sensors Moraveji is wearing are part of a system called Calm Coach. Put it on in the morning when you start work at your computer, and over the course of the day, it monitors your breathing rate.

Moraveji points to an indicator on the Mac's menu bar that shows his current breathing rate and compares it to a baseline. The numbers are low. Unlike most graduate students, he does not feel stressed when talking about his PhD research.

Moraveji points to another number on the menu bar. "We're trying to represent something ephemeral but important," he says, "your state of calm. We do that with these points." Calm Coach rewards you with points for breathing well. (As I look, the 37 beside his heart rate switches to a 38.) It doesn't deduct points if the rate goes up during an angry call or a bad meeting. As any gamer will tell you, losing points makes users even more stressed. Besides, it can be useful to see when in the day you're stressed and what you're doing when that happens. Moraveji calls up a page with two columns and a set of screen-shots of his laptop. "The column on the left shows what I was doing when I was most stressed, and the one on the right shows what I was doing when I was most calm," he explains. (I notice that in two of the screens on the left, his e-mail is open. I feel a little better that even experts on calming sometimes have trouble remembering to breathe when they're checking their mail.) Using this system for several weeks or months, someone can see what times of day he's most likely to be stressed, as well as what activities he handles with greatest ease.

A picture of a beach pops up on the screen. "I hit a milestone," Moraveji says; the system is rewarding him for reaching 41 points that morning. Future versions of Calm Coach, he says, could be even more proactive: they might suggest you take a break when your stress level gets too high or suggest you spend the hour you're normally at your calmest doing your most challenging work.

Calm Coach still has the charming rough edges of a prototype: the Arduino controller doesn't have a case (troubleshooting is easier without one), and a wireless breath monitor would make it a lot more usable. Still, I can imagine people who already wear heart monitors when they exercise or who are always on the lookout for tools to boost their productivity using a sleeker, more polished version. The real beauty of Calm Coach is that it's always on, always collecting data, always providing bits of feedback and nudging you to be calm. Given that the average adult takes more than twenty thousand breaths a day, the virtue of having the system working whenever you're at the computer seem obvious. A system that's real-time yet largely indirect seems well suited to help someone reprogram the largely unconscious connection between technology use and breathing.

But while it might sell first to the lifehacker crowd—high-tech early adopters who've already bought the Nike+ FuelBand; have dog-eared their copies of *Getting Things Done;* and are hunting for the next thing to make themselves smarter, more productive, or more *something*—Moraveji hopes that the tangible benefits will nudge all users to a greater interest in mindfulness. "People who don't want to be contemplative but still want to reduce their stress or get their work done can benefit," he says, "but really, it's about increasing emotional and physical awareness." After all, "calm isn't just about soothing. Calm is about quieting the mind so you can be productive and creative, to have great ideas." He also hopes that Calm Coach and similar systems can help show users that "the technologies that stress us out will calm us down," that we can hack and rewire both our computers and our relationships with them. "A computer shouldn't just enable me to *do* things," Moraveji says. "It should help me be the best me I can be."

Calm Coach is still some way from hitting store shelves. Until it's available, there are other tools that can help you learn to be more contemplative with computers. They go by the promising name of Zenware.

TWO

SIMPLIFY

The next time you're at your computer, download two pieces of software. One is Freedom, a program than blocks your access to the Internet for up to eight hours. The other is Dark Room (or WriteRoom, for Macintosh users), a writing program with a clean, simple interface designed to help you focus. (If you use Linux, you're enough of a hacker to find your own versions of these.) Spend a week with these programs, and they may start to improve your writing and concentration. They may also help you learn a couple things about yourself. Contemplative computing requires experimentation and reflection; it's important to try new things, see how they affect your extended mind, and change your technologies to help you develop that mind and support your ability to be creative and focused.

Freedom is very easy to use. When you open it, it presents you with a dialogue box. *How many minutes of freedom would you like?* it asks. Type in a number, hit Return, and you're offline. Nothing you can do gets the Internet back while the clock is counting down. If you want to get your e-mail or Twitter fix, you'll have to restart your computer. And if you do that, you get to ask yourself, *Why am I doing this?* It's a surprisingly effective deterrent.

The first time the program asked me *How many minutes of freedom would you like?*, my initial reaction was mild panic. *Not connected? What am I doing? Am I crazy?* Living with the Internet has created a reflexive need to be connected, even when we think we would benefit from being offline, and the assumption that we *have* to be online is the first brick in a road that leads to cat videos and hair-trigger e-mail checks. But of course I have my iPod and iPad if I need to be online. It's not like I'm sending myself to Siberia.

Still, I quickly check my mail and back up my book files to my server before I give myself two hours without the Internet. I hit the last button, and Freedom announces, *You are now offline. Freedom will not respond until your offline session expires.* After a minute or two of writing, I command-tab over to Freedom. It's completely unresponsive: no menu bar, no *Get back to work,* nothing. I'm reminded of the scene in *Young Frankenstein* in which Gene Wilder's Dr. Frankenstein (*I wonder if that clip is on YouTube? Oh, damn, I can't look*) tells Teri Garr and Marty Feldman (*There was that Sherlock Holmes movie he and Wilder did; does Netflix have—no, can't look*) not to come into the room with the monster no matter what bloodcurdling screams they hear (*That reminds me, I—no, forget it*).

It amazes me how often during a single (admittedly rather trivial) thought my mind wants to veer off onto these other paths, pick up this idea and that one, answer this or that question—and how easy the Web lets me satisfy that curiosity. What makes the Web so damn distracting is that I probably *could* find that clip, the movie could well be on Netflix, and between it and the Internet Movie Database, I could waste several minutes wandering through the back roads of Marty Feldman's tragically brief career. We fail with the Web because it succeeds. Or, more accurately, we fail to stay focused because we know that every second there's an amusement park–size virtual world where we can distract ourselves.

But unless I restart my computer, I have no choice but to settle in and actually write. Every now and then I think, *I wonder if I've got any new e-mail,* or *Has Felix Salmon said anything smart about the latest monetary crisis?* And that's all I can do: wonder. Not find out. I can't. I've turned off the Internet. So I go back to work.

After a while, those thoughts turn from *Damn, I can't check my mail* to *Cool, I can't check my mail.* It really does feel like freedom.

Freedom shows us that when we need to focus, less can be more. WriteRoom does the same thing. When you first open it, WriteRoom takes over your entire screen; all you'll see is a blinking green cursor against a black background. There are no menu bars, no fonts to adjust, no IM windows peeking around the edges, no status updates or e-mail alerts beckoning from the periphery. If your computer looked any more peaceful, it would be turned off, but the presence of the cursor conveys a sense of the machine's readiness to be useful.

To users accustomed to a constant peripheral buzz of infinite drop-down menus, likes, ratings, instant notifications, and real-time recommendations, this will seem punishingly stark, like a prison cell or a room that's been looted.

For me, though, its visual minimalism evokes memories of a time when personal computers seemed full of possibility. Even if I couldn't really make them do very much, and what I could do usually required hand-entering programs in BASIC, I knew that the Apple II, Commodore 64, and TRS-80 were the first pieces of a universe already mapped out in science fiction, just waiting to be assembled.

I'm not the only one to react to WriteRoom this way. If you were a teenager when the first PCs came out, in the late 1970s and 1980s, the program has the potential to trigger old cybernetic memories, a Möbius strip–like nostalgia for a future that was about to happen. Virginia Heffernan raved about WriteRoom in terms that were part Virginia Woolf, part cyberpunk: "You rocket out into the unknown, into profound solitude, and every word of yours becomes the kind of outer-space skywriting that opens *Star Wars,*" she writes. "For those of us who learned Basic on a Zenith Z19 and started word processing on a Kaypro (anyone?), the retro green-and-black now takes the breath away."

It was a time, Heffernan says, when "the mystery of the human mind and the mystery of computation seemed both to illuminate and to deepen each other." Computers were new worlds waiting to be explored and mastered. Sitting down in front of the keyboard wasn't something passive and inactive. It was the beginning of an adventure, a personal quest. That exploration promised to change you, to make you smarter, to give you control over this little world (if you worked enough to unlock its secrets). In the earliest years of personal computers, when you had to write (or at least type in) your own programs, it was clearer that you and your computer made each other smarter.

But today, computers are more likely to feel like they're incomprehensibly fast and complex, and software that's meant for everyday use can be dizzyingly complicated. My copy of Microsoft Word, for example, has eleven drop-down tabs in its menu bar. Spread out among ten of those tabs are a hundred and forty commands and features, ranging from Open and Save to Flag for Follow-Up and AutoFit and Distribute. The eleventh tab has over two hundred fonts, not counting the many bold, italic, underlined, condensed, and light versions of each. Just in case drop-down menus aren't your thing, you can access some of the functions graphically with a ribbon that runs across the top of your document. You can get to others by using a toolbox that sits to the side of the screen, and yet others with a sidebar on the left of your document. Finally, there are six different ways to view my document: draft, outline, publishing, print layout, notebook layout, and full-screen.

Programs like Word give users fewer chances to find what Heffernan calls the "illuminating mysteries" that simpler, barer systems offered. Even technologies designed to make us safer or more productive—"user-friendly interfaces" intended to "protect us from technology's dark places," as Heffernan puts it—run the risk of unintentionally dulling our skills or intuition. This doesn't happen just at the keyboard, and in the real world, the consequences can be deadly. A 2011 report by the International Air Transport Association, issued in the wake of the 2009 crash of an Air France Airbus off the coast of Brazil, warned that airplanes were becoming so sophisticated that pilots didn't have the chance to develop and maintain advanced flying skills. They're trained to leave planes on autopilot for most of a flight, so they have fewer hours' experience flying manually and therefore have a harder time handling emergencies, especially when the emergencies are caused by problems with the autopilot or instruments.

So we need to work out for ourselves which quicker or more complex software really makes our lives better. It's easy to assume that as computers get faster and more powerful, we get faster and more powerful too. But decades of studies of complex systems failures show that highly automated systems can never completely eliminate the underlying complexities of the world—the intractability of the laws of physics and the unpredictability of the weather. And these are the same systems that make it less likely that pilots (or nuclear power-plant operators, or hedge-fund managers) will have the firsthand experience with those underlying complexities that's necessary in order for them to assess and act with the kind of disciplined, well-trained calm that you want when things go sideways.

If you're a scientist or financial analyst and you work with very large data sets or simulations, there are tasks that would have taken you days to do before computers that are trivial now, and there are other things you can do now that would have been impossible without computers. In certain fields, software has been like electricity or engines: it's clearly made work faster. But not all creative work can be supercharged by software. Consider writing. In the twenty years since I started using Microsoft Word, there've been ten turns of Moore's law. Moore's law essentially states that roughly every two years, the power of computers doubles. That means the computer I'm using now is about a thousand times more powerful than the one I used to run Word 5.1. Am I writing a thousand times faster? Are any of us? Are we able to read a thousand times as much e-mail as we did in 1990? Or does it just feel like we get a thousand times as much?

Big, feature-rich programs are complicated. Creative work is also complicated. They're just complicated in different ways. Radically simple programs support your efforts to deal with that complexity by stripping away peripheral distractions, blocking the outside world, and giving you the space to think and multitask more effectively.

Multitasking has gotten a bad rap lately, but people have been multitasking forever. Literally. According to some archaeologists, *Homo sapiens'* success is a function of mankind's capacity to multitask. One of the most fascinating examples of ancient multitasking comes from experiments conducted by Lyn Wadley and colleagues at the University of the Witwatersrand, in South Africa, in which they reproduced methods of making ancient stone tools. In the Stone Age, hunters made axes and other weapons by attaching stone blades to wooden handles. Securing these pieces together, a process called hafting, required strong adhesives, which the hunters made by gathering natural ingredients and cooking them. Learning to do this well would have required a lot of experience and skill. Wadley and her colleagues argued that there was another critical ingredient in making such adhesives: the ability to multitask.

To understand their argument, it helps to think back to chemistry class. In a typical experiment, you would mix together chemicals in the proper proportions and the right sequence, then heat them for *x* minutes at *y* temperature, maybe testing the pH, perhaps stirring at a critical moment or adding another ingredient when the compound changed color but before it changed *too* much. Even under controlled laboratory conditions, this could be a challenge. Now imagine doing all that in the outdoors. You have no pure ingredients; you are working with components from things you've foraged, killed, plucked, grown, or dug. You can't measure *x* and *y* exactly, because you've never heard of the concept of standardized weights and measures, and maybe not even the concept of numbers. And you have to do everything while watching the fire to ensure it doesn't go out, get too hot, or become too uneven. What kind of grade would you have gotten?

These are the conditions Stone Age craftsmen worked in, and they weren't graded on a curve; there were only two grades—life and death. Ancient peoples had used natural adhesives like tar and tree sap for a long time before this, but while they worked for simple things, they weren't strong enough for use on weapons. About seventy thousand years ago, ax-making peoples in South Africa discovered that a blend of plant gum and red ocher (an iron-rich mineral) could make an adhesive strong enough to keep an ax head attached to its handle for years—but it worked only with the right plants and the right mix that was heated for just the right amount of time. Wadley's team attempted to reproduce the glue and found that trace amounts of beeswax and grains of sand added to the gum-and-ocher mix stabilized the adhesive if it was too runny or thick; they also found that even in the lab, it was easy to put in too much and ruin everything.

So, to be successful, Stone Age adhesive makers had to know how their ingredients would perform when heated, and they had to determine the optimal level of heat for each batch and decide whether and exactly how much additive was required. In other words, they needed to be able to predict the performance of different ingredients, watch their brews constantly, and make informed adjustments to the recipe to avoid disaster—and waste. All this prediction, observation, and improvisation, Wadley concluded, would have been "impossible without multitasking and abstract thought."

On the other side of the world, UCLA professor Monica Smith also found evidence of multitasking in the archaeological record. Indeed, she too argues that man's capacity for multitasking was an important factor in the evolution of complex societies. In our species' early history, *Homo sapiens* competed for food and space with a number of stronger, quicker, or more aggressive species (some of which, like pigs and cattle, were later domesticated). The ability to multitask, at both an individual and a group level, allowed our ancestors to make up for their deficiencies. As individuals, they could plan and execute more complex tasks than other animals and expand their range of useful food by burning, curing, or cooking it. At the group level, they could move from gathering nuts, fruits, and tubers to hunting and growing food (an enterprise that requires lots of planning, as well as the ability to invest energy now for bigger payoffs in the future). Multitasking let our species create more complex social rituals and technologies and, eventually, a more sedentary and urban way of life. In other words, multitasking helped our ancestors succeed against other species, build tools, and ultimately create civilization.

So if humans have been multitasking for tens of thousands of years, what's wrong with me texting my kids and sipping my iced decaf triple-mocha latte while speeding down Highway 101? (Everyone else is doing it.) If multitasking was so critical to human social evolution, why stop now?

The simple answer is that we use the word *multitasking* to describe two completely different kinds of activities. Some are productive, intellectually engaging, and make us feel good. Others are unproductive, distracting, and make us feel stretched thin. It's important to distinguish these different kinds of multitasking, because we use the word very casually, and often incorrectly—and a lot of what we call multitasking is something else.

The good multitasking is Stone Age multitasking. It's the kind that you do when you're doing something that completely engages you.

For Lyn Wadley, multitasking is the ability to "hold many courses of action in the mind." It involves the capacity for abstract thinking and the ability to switch one's attention between different objects or parts of a process. Monica Smith defines multitasking as "the ability to do more than one activity at a time, and the ability to adjust the timing and sequence of activities in response to changes brought by external or internal conditions." In both cases, multiple courses of action converge on a single end point: making an adhesive suitable for hafting; preparing raw materials for a meal; or preparing a field for planting.

This is the kind of multitasking that we do when we're engaged in complicated projects; we have to juggle lots of balls so they'll land in the right places, in the right order, at the right time. We do this fairly naturally, and if Wadley and Smith are right, we've evolved to do it.

There are other circumstances in which we can constructively use different cognitive strands. A good lecturer will use a few words on the whiteboard or an image on the screen to help students remember the key points she makes. Among graphic recorders—people who draw remarkably elaborate maps of talks or conversations—it's an article of faith that watching a topic map evolve while people are speaking reinforces learning. In these cases, different cognitive streams don't compete with one another; they reinforce one another, delivering similar messages in different ways. Watching an opera requires multitasking: your brain has to process the music, plot, lyrics, and staging and put them all together to absorb the rich message.

Cooking offers as good an example of multitasking now as it did when mammoth was on the menu. Imagine you're making dinner for friends. You need to think about what your friends will eat, then plan your ingredients and decide when each should be acquired (cornmeal and beans last longer than fish and fresh basil). Each dish has to be prepared so everything is ready in the right order and so that cooking one doesn't interfere with another. While you're making the meal, you'll probably have to take care of other tasks too, like ensuring that the dishes are clean and that the tablecloth and candles are ready. You may have to adjust your schedule after hearing that two guests are stuck in traffic—the roast will be fine, but the pie will go in the oven later, and the kids should put out more crackers now and whip the cream later. Is it a challenge to keep all this on track? Sure. Is it rewarding when it all works, when guests think everything is going smoothly? You bet.

This capacity for braiding together different, convergent cognitive strands is one we can also apply to purely intellectual activities. Without the ability to hold a couple different ideas in short-term memory, we would find it extremely hard to compare them, to see connections between them, or to forge new ideas. Lots of creative works and innovations involve surprising combinations, the juxtaposing of familiar things in unfamiliar ways that makes something new. That kind of creativity would be impossible without a capacity for multitasking.

People also use the term *multitasking* to refer to two activities that can be done at the same time because one is very familiar or because they involve different parts of the brain. Lots of people listen to music while reading or writing (some interesting research suggests that a person's psychological profile determines whether he prefers instrumental or vocal music), think deeply while walking the dog, or talk on the phone while keeping an eye on the baby. It's great that I can fold laundry and listen to music at the same time, because I'd never do it, or any other housework, if I couldn't. Indeed, mixing these kinds of activities barely counts as multitasking, though some think of it that way; I once had an elderly swimmer compliment me on multitasking when I was sitting in a Jacuzzi and reading. (Of course.)

There are times when it can get hectic, but it's still well within a person's abilities to cook three different dishes on the stove and in the oven while getting the kids to set the table. Reading at a desk crowded with books and monitors while struggling to braid together several ideas is a challenge, but one that's absorbing and rewarding. These kinds of multitasking encourage flow.

But the multitasking we do when we split our attention among several devices or media is something else entirely. Writing at my desk with music playing is very different from tabbing between two Web pages in my browser while also chatting with a high school friend on Facebook and listening to a podcast on my iPhone. These separate activities don't add up to a single grand intellectual challenge. They're just different things I try to do simultaneously.

This kind of multitasking, scientists will tell you, isn't actually multitasking, despite the label. It's switch-tasking: your brain is toggling between different activities, constantly redirecting its focus, tearing away from one task to deal with another.

Why is switch-tasking a problem? In addition to making us less creative and productive, addicting us to inefficiency, and causing us to be more susceptible to self-delusion, switch-tasking is a lot harder for the brain to manage than we once thought. This became clear when I met a UC-Berkeley psychologist studying memorization and multitasking.

Megan Jones and I meet at a café across the street from the university and find a table after getting our lattes. (Science on the West Coast can be a bit laid-back.) She gets out her iPhone and turns on the stopwatch app. "Okay, we're going to do a simple three-step experiment," she says. "First, I want you to count from one to ten as quickly as you can."

I feel a tad self-conscious doing something so basic around people who are reading heavy tomes and textbooks, but I do it, counting quickly and softly so as not to disturb the woman next to us who's immersed in *Middlemarch*. "That took about a second and a half,"

Megan says when I finish. "Okay, the second step is to recite the alphabet from A to J."

Easy enough. Again, it takes about a second and a half. The woman reading *Middlemarch* conspicuously ignores the idiot beside her. She turns a page.

"Now," Megan continues, "try alternating them: go one A, two B, and so on, up to ten J." If I did each of them in a second and a half, I figured, it should take three, maybe four seconds to do them together.

I start: "One A." The first several letters and numbers are easy, but around "five E," I become intensely conscious of the fact that the numbers and letters no longer come automatically. I now have to *think* about what I'm doing. I start saying them louder to keep focused. "Umm, six... F, seven..."—*E, F, next letter is*—"G..." *What number did I just say? Oh*—*Middlemarch* looks up, which completely flusters me. "Eight..." *Damn, damn, damn.*

I trip through the rest. "J" follows "ten" quickly, but only because I know that I'm at the end of the sequence. "That was nine and a half seconds," Megan says. I try a couple more times, but even my latte-charged brain can't get from 1 A to 10 J in less than nine seconds. Distraction over, *Middlemarch* returns her attention to her book.

This is a classic experiment on switch-tasking; it's practical, and not only because you can do it in a café instead of a lab. Everyone has recited the alphabet a million times, and we can all count to ten in our sleep. You barely need to pay attention to the first two parts of the experiment.

Mix the two tasks together, though, and they suddenly become nonautomatic. Having to think about the next number *and* letter slows you down dramatically, throwing a normally smooth mental operation completely out of gear. In my case, it took three times as long to complete the tasks when they were interwoven—in other words, when I was switch-tasking.

Now imagine trying to listen to a conversation about one subject while writing an e-mail about another, or trying to participate in a meeting while scanning the headlines. Or don't imagine it; just remember the last time you did it.

It may feel like dividing your attention between two simple tasks is a relatively easy thing, but the café experiment reveals that switch-tasking is expensive. Every time you move from one window to another on your computer or move from reading your e-mail to listening to a conference call, your mind has to spend energy. By some estimates, you can lose several working hours every week to these moments, which come at exactly those times when you want to be most productive. You also make more mistakes when you switch-task.

Studies have shown that switch-tasking can be dangerous. When you're driving and you answer your cell phone, part of your attention switches from monitoring your environment to listening to the conversation, and you're less able to notice and react to sudden, unexpected changes in traffic—the car that switches lanes without signaling, the child who runs out into the street. Even after a call is over, it takes a few seconds for you to shift from focusing on the phone call to noticing what the cars in front of you are doing.

You're also less creative when you switch-task. It's easy to believe that multitasking raises the odds of making novel associations between ideas, and it does—when the different activities are directed toward a single goal. Switch-tasking, however, emphatically does not make you more creative. During switch-tasking, your brain spends so much energy doing basic management that you have little bandwidth left for seeing previously invisible connections or making new associations.

The irony is that switch-tasking is self-protecting. Stanford professor Clifford Nass has found that most dedicated multitaskers (really switch-taskers) are actually "terrible at every aspect of multitasking. They're terrible at ignoring irrelevant information; they're terrible at keeping information in their head nicely and neatly organized; and they're terrible at switching from one task to another." Yet, in a sad twist, compulsive switch-taskers think they do it well. There's something about switch-tasking that makes one overestimate his own competence and downplay the costs.

Unfortunately, devices like cell phones are tailor-made for switch-tasking. They capture your attention by diverting it from something else. Digitally enabled switch-tasking tends to push several tasks into a narrower band of attention in a way that seems to short-circuit your ability to really focus when you need to. Participating in a meeting while writing an e-mail to a friend doesn't activate different parts of your mind; both tasks compete for attention from the same part of your brain. And some troubling data suggests that people who are heavy digital switch-taskers have a harder time than others concentrating for long periods. Once the brain becomes accustomed to having multiple inputs and distractions, it's difficult for it to settle down and do a single complex task.

The cognitive resources required to switch-task are vastly different from the resources required to read lots of different things as you prepare a speech, or to keep track of three dishes on the stove as you prepare dinner. At a personal level, you experience switch-tasking very differently than you experience multitasking to reach a single goal. Mixing the gum and ocher, watching it heat, stirring it, getting a feel for how it's cooking and what you need to do to it next creates something close to a flow experience, not the sense of having your attention pulled in several different directions. As Nass says, our ancestors' early environments provided lots of challenges and stimuli, but they probably experienced it all as interconnected. "If you were out hunting an animal," he says, "you might look at a lot of things, but they were all about hunting that animal."

While writing this section, I've had three books and my journal open on my desk, and a bunch of browser tabs and a PDF of a scientific journal open on my iPad. I can move between them as I research one thing and then another, check citations, look up references, and make sense of what I'm reading. I'm doing a number of things, but unlike those times when I'm talking on the phone to a colleague while trying to respond to a child's question, these activities are organized around a single purpose: understanding the history of multitasking. I also have the Band's *Music from Big Pink* playing in the background. For me, music is a source of psychic energy, and it helps me concentrate. I can tune out lyrics without too much trouble, though spoken words interfere with my ability to write (so no rap or news podcasts). As I write, I'm barely aware of the physical patterns that my fingers trace across the keyboard to form words; because I spell with my hands, I don't have to be conscious of the combinations of letters that make up words. I can think about sentences and arguments.

Writing itself actually incorporates a variety of cognitive activities and illustrates how in the course of pursuing creative goals we become entangled with our tools. That complexity is one reason writing requires intense focus, and it's why authors notoriously seek solitude to write. (A number of famous authors and thinkers did good work in the austere environment of a prison cell; Marco Polo, the Marquis de Sade, Oscar Wilde, Saint Paul, Niccolò Machiavelli, Cervantes, Ezra Pound, Aleksandr Solzhenitsyn, Mahatma Gandhi, Antonio Gramsci, and Martin Luther King Jr. all produced some of their most memorable works while incarcerated. Throw in prisoner-of-war camps, and the list expands to include French Annales historian Fernand Braudel and philosopher Ludwig Wittgenstein.) So it should come as no surprise that some of the best examples of Zenware are found in writing programs.

Zenware (the term was coined by freelance journalist Jeffrey MacIntyre in 2008) begins with the premise that simple tools are more valuable than complex ones when you're multitasking. The writing programs stay simple in order to keep an already-complicated and already-challenging task from becoming even harder. Zenware doesn't try to "solve" the problem of writing's difficulty by automating tasks, and it doesn't try to maximize your productivity by offering lots of features. If Microsoft Word is a fly-by-wire Airbus 380, Zenware is a DC-3. It disrupts users' existing work habits as little as possible, forces them to learn few new commands and concepts, and minimizes the cognitive load of the software itself. It eliminates external distractions, leaving more of the mind free to multitask productively.

Hog Bay Software's WriteRoom, a program that helped define Zenware, is a case in point. As developer Jesse Grosjean recalls, WriteRoom didn't start simple; it was at first a full-screen mode for a much larger and more complex outlining program, a type of software popular with book writers, that he was coding. He soon found the full-screen editor more compelling than his outliner, and after a solid week's programming, he had a working version of WriteRoom. It spawned numerous imitators, including Dark Room and PyRoom (versions for Windows and Linux operating systems, respectively) and helped make minimalism a design feature.

What made WriteRoom distinctive? Partly it was the program's underlying philosophy, its approach to writing and attention. For decades, word processors have been guided by the principle of WYSIWYG, or What You See Is What You Get—that is, what you see on the screen should look as much like the finished product as possible. As computers became more powerful, developers were able to give authors more and more control over What You See: the first word processors had only a handful of fonts, for example, while today's can have hundreds—even thousands, if you write in more than one language. Along the way, though, the user's ever-growing ability to endlessly tweak the look and feel of the document started to intrude on writing. Changing margins, column sizes, line spacing, and layouts have become the digital equivalent of tasks like sharpening pencils and cleaning out desk drawers, activities that seem like work but are actually ways to avoid work.

Outlining programs try to help authors manage large, complex projects by being large and complex themselves. A program like Scrivener, for example, encourages writers to create many small text documents rather than big chapter-size ones, to organize the documents into chapters using an outline tool, and then to place the chapters (as well as primary documents, Web links, and the like) into binders. Documents can have labels, tags, notes, comments, and lots of other metadata. The program also has a number of other tools, like a manager that shows how many words you've written so far (or today) and how many more you need to produce before you're finished. Scrivener is designed to help the author see the structure of his book so he can identify sections that need to be written and reorganize the chapters at will. But it has a steep learning curve. After using it every day for a year, I was still discovering new features.

WriteRoom takes things in the opposite direction, confidently proclaiming that when it comes to writing, less is more. Grosjean described WriteRoom as a "writing environment" rather than a word processor or text editor. It has none of the features that software writers expect in a text editor (syntax highlighting, for example), nor does it do document styling and structuring like Word. "It doesn't provide features as much as *feeling*," he said. WriteRoom focuses relentlessly—and focuses *you*—on the moment when words go on the page, and it devotes no resources to formatting or printing.

Reviewers and users quickly came to understand and appreciate the creative value of Zenware's simplicity. "This type of word processor is about the process of writing itself," Canadian author and Web developer Michael Gorman explained. "It is not about making your words look pretty, making tables, or about adjusting fonts and sizes." Indonesian programmer Donald Latumahina reported, "The fact that there's nothing else displayed on the screen allows me to focus entirely on the task in front of me. No disruption, no eye candy, nothing. Just me and the task." Another user said, "When you're in it, you're in it. You can't see your system tray, your Start button, your desktop or anything else. Just the editor. It's fantastic for getting things done." German author Richard Norden wrote, "All there has to be is me, a blank screen, my words and the current word count. No fancy toolbars, colorful buttons, floating windows or other useless distractions from the things that really count."

A second widely imitated feature of WriteRoom is its look. WriteRoom doesn't look like paper, with black letters on a white background; it has a black background with bright letters, like a late 1970s mainframe terminal. As the *Washington Post*'s Rob Pegoraro put it, "WriteRoom seems to have cast a shiny new Mac laptop back to the dark ages of DOS." But in an era when "programs sometimes throw information at us, rather than help us process it," Pegoraro continues, the minimalism of the new apps wasn't just "curmudgeonly" or "computing nostalgia." This retro aesthetic was "a great way to get the attention of early adopters," Grosjean said. Given that such people usually demand bleeding-edge design, this might seem counterintuitive, but you can see a similar impulse at work when electronics companies release digital cameras soaked in the aesthetic of the Leica M3, a classic design that first appeared in the 1950s and that for decades set the standard for simple, serious photography, or in an app like Instagram, which makes new photos look old and somewhat simpler, but richer.

If WriteRoom's minimalism is black-screen-and-blinking-cursor old school, OmmWriter Dana, another piece of Zenware, is the kind of minimalism you might see in a boutique hotel in Helsinki. It's a masterpiece of thoughtful emptiness, a few compelling elements surrounded by negative space.

The fact that it's the brainchild of an advertising agency makes it even more intriguing.

Marzban Cooper, one of the principals at the Barcelona-based Herraiz Soto, explains that the company got into software development because of "a real need from inside the agency." As a small firm that designs Web sites that "turn consumers into fans," its writers have to create focused, compelling copy that draws in visitors. But in the perpetually chaotic world of advertising, it can be hard to focus. Agency cofounder Rafa Soto came up with the idea for OmmWriter on vacation, sitting "on a deserted beach in Brazil." When he returned to the agency's offices, Rafa assembled a small team of six designers and programmers and got to work. Fourteen months later, all of the agency's writers shut down Pages and Word and opened OmmWriter for the first time.

Opened, in this case, is exactly the right word. You don't just write with OmmWriter; you put yourself in it. *Welcome back to concentrating,* it says as it turns off e-mail and chat notifications. OmmWriter fills your screen with one of three backgrounds; your words can hover over a snowy landscape with bare trees under a gray sky, in front of a simple white canvas, or on a blue-gray background.

The program was gently influenced by Soto and Cooper's experience with meditation, and it acquired a "Zen look and feel, tone of communication and philosophy," Cooper says. But as the team worked to create something with the simplicity of "a pen and paper and nothing else," OmmWriter moved farther away from conventional programs. It was "definitely not a word processor," Cooper adds, since it intentionally lacked the features that make finished documents look good but distract writers at the creative stage. And "to call OmmWriter a text editor is to belittle it," Cooper notes.

OmmWriter evolved instead into "a sanctuary, a space for you to be alone with yourself and your thoughts." (Herraiz Soto's Barcelona office is that open style beloved by companies from Silicon Valley to Tokyo for its ability to facilitate collaboration, sometimes at the expense of concentration.) And indeed, it has a kind of enveloping quality, thanks to what's not there—pop-ups, controls, and options (the menu consists of discreet buttons that only appear when you move the mouse)—and what happens when you use it. As you type, the letters make a splashy computer terminal–like click, as if digital water were flowing over the keys.

True, there are only a couple fonts; however, it's the only writing program that comes with its own playlist, a set of tunes scored by Barcelona sound engineer and ambient composer David Ummmo. "It's like Brian Eno made a word processor," one user commented.

When you close the program, it tells you that notifications are restarting and adds the Zen-like and slightly worrying *Your mind, a wild monkey*. Seeing that, one is tempted to start it up again immediately.

After its introduction within Herraiz Soto, OmmWriter caught on among the firm's copywriters. No one uninstalled Word or Adobe Creative Suite—everyone still needed those programs to turn words into Web sites—but Cooper noticed that people "started to use OmmWriter not only for writing but for thinking, for finding unique concepts and ideas." At this point, Soto and Cooper decided it "was just too good to keep to ourselves." They renamed it OmmWriter Dana and posted a copy of it online in November 2009; an iPad version appeared in May 2011.

Since 2009, OmmWriter has been downloaded by several hundred thousand people, and it has "opened doors to clients that we previously would never have reached, especially internationally," Cooper says, adding that "OmmWriter is now Herraiz Soto's most important calling card." For the technically sophisticated agency, the program's attention to detail, the color-therapist-selected palette ("one which stimulates creativity, and the other which promotes tranquillity"), the background pictures commissioned from a British photographer, and, especially, the beautiful minimalism show off the firm's technical chops and design style.

OmmWriter can be so distinctive because Zenware is very auteur-driven. Most software today has all the uniqueness of a public ad campaign; a company's suite of programs will share the same design sense, common icons, color palette, and so on. Operating systems impose their own rules, and developers are encouraged to mimic the look and feel of the latest version (remember the rage for aqua icons?). WriteRoom, OmmWriter, and other programs, by contrast, represent the aesthetic vision of individuals, and many were inspired by a specific need or an aha moment. As software developer Jesse Grosjean explains, "A simple text editor is probably the easiest program to write." Unlike many other kinds of software, it can be developed and maintained by a single programmer or a small group. The entire global Zenware-development community might have enough people to field a single softball team; Microsoft's hundred-person-strong Word group, which is updating a twenty-year-old program, could make its own softball league. But while the small number of developers working on even the most popular Zenware may reinforce a preference for ruthless simplicity—you can't bloat your program with extras when you also have a day job—the drive to minimalism doesn't constrain the possibility of personal expression. For all their simplicity, these programs are all different from one another, in the way that the writing styles of Ernest Hemingway, Raymond Chandler, and Jeanette Winterson are all different. Each is spare but still unmistakable.

Zenware isn't just writing tools. There are Zenware programs that pull you back from the world of online ads, games, updates, and real-time information and clear a space for you to recenter your attention.

The simplest examples just tweak your computer's normal interface, highlighting the software you're working on and obscuring other programs. In effect, they keep a virtual spotlight fixed on the programs you're using at the moment, but they do so in different ways. Backdrop and Think allow you to have multiple programs open, but they keep only the program you're working on visible; everything else is placed behind a virtual blank screen. HazeOver and Isolator take a less extreme approach by dimming, not disappearing, programs or windows that are open but not currently in use. Finally, Shoo Apps reflects your own concentration back to you by highlighting the programs you've used in just the previous couple of minutes.

These programs try to strike a balance between focus and switch-tasking. The visual foregrounding of one program keeps the others in the periphery of your awareness. The programs are designed to help you stay focused on the most important of several tasks, not to reduce the number of things you're doing.

Other programs take a different approach to helping you focus. They break your Internet connection.

Some are browser plug-ins that keep you from visiting time-wasting Web sites. Chrome Nanny and StayFocusd (complete with hip, Web 2.0–compliant misspelling) work from a list of Web sites you want blocked and either deny you access to those sites during certain hours or allow you to visit them for only a certain number of minutes per day. Stand-alone programs can present more formidable defenses to online distraction. SelfControl lets you turn off e-mail and denies you access to blacklisted sites. Antisocial limits your access to social-media sites. Freedom just shuts everything down. (It's "a very brute-force attack on the problem of connectivity," its creator admits.) Freedom and Self-Control won't unblock your access even if you turn them off. In fact, SelfControl won't let go of your Internet connection even if you reboot your computer. In the world of productivity software, it's the honey badger.

The number of these programs and the variety of approaches they take to nudging users to stay focused suggest how pervasive the problem of online distraction really is. So do the backgrounds of their developers. Think is the brainchild of Freeverse, a New York software company best known for strategy games and SimStapler, a quirky office-equipment simulator that "brings all of the thrill and excitement of a 'real' stapler" right to your computer screen, according to the app's description. StayFocusd is the product of Los Angeles digital advertising studio Transfusion Media. SelfControl is the brainchild of Steve Lambert, an artist and activist whose other credits include an elaborate fake *New York Times* issue in November 2008 that announced an end to the Iraq war and an app that replaces Web advertisements with art. He designed SelfControl because "as anyone who creates things knows, the time you can block out to get focused work done is invaluable."

If you want to see how the Internet can distract even the most creative people and how they fight back, you couldn't do better than to look to LeechBlock creator James Anderson. Anderson holds two PhDs from the University of Edinburgh, in computer science and philosophical theology. He's probably the only person who'll ever publish articles in both the *International Journal of Human-Computer Studies* and the *Calvin Theological Journal*. Anderson spent thirteen years at the Centre for Communication Interface Research at Edinburgh, and for the past several years he has been a professor of theology in North Carolina. But even a computer-scientist-turned-theologian can spend too much time watching YouTube videos and falling down "Wikipedia rabbit-holes" at work, it turns out. "I realized I needed to take drastic action," he says.

For a man with a computer science PhD, drastic action could mean only one thing: writing code. There were already scripts that blocked Web sites, but Anderson needed finer controls—with his research projects on things like electronic commerce and interface design, he couldn't afford to go offline completely. He designed LeechBlock to limit his access to time-wasting sites. You might block Stuff White

People Like entirely, allow yourself ten minutes on Facebook after work, and allocate an hour during work for Web sites like Digg and Slashdot, sites that mix useful, work-related content with frivolous stuff.

It took only a couple of hours to write the first version, he says, but "I've spent many more hours developing it since then." The program made a huge difference to his own productivity, and Anderson says he "figured that I probably wasn't the only person with that problem. So I uploaded the extension."

Anderson sent LeechBlock 0.1 to Mozilla's online repository of Firefox browser extensions in February 2007, the same year his monograph on paradox in Christian theology was published. It might seem odd that a program that makes *less* of the Web accessible could qualify as a Web browser extension, but the success of LeechBlock illustrates how *functionality* can mean very different things for computers and for people. Three years, twenty-two updates, and "probably hundreds of hours" spent in development and documentation later ("Yikes!" he says after adding it all up), LeechBlock has been downloaded more than half a million times.

Anderson isn't a technologist-turned-Luddite. When I ask him whether computer science and theology have anything in common, he replies, "Actually, they have more in common than most people imagine at first. I'm very left-brained, so in a sense, all that changed was the subject matter to which I was applying my style of thinking." For him, and for the fifty thousand people who use it every day, LeechBlock isn't about rejecting computers. It's about making them work better.

Fred Stutzman, the developer of Freedom, isn't a Luddite either. "My dad brought home an early PC XT when I was ten," Stutzman recalls, and as a teen he taught himself BASIC, learned his way around DOS, then spent the next decade using Linux. He seems like a typical programmer, but he doesn't have a typical programmer's attitude toward people. He's more "interested in how technologies affect social practice and how they affect how we interact with each other" than in technology itself, he explains from his office at Carnegie Mellon University, where he's a postdoctoral fellow studying digital privacy.

The idea for Freedom came to him in a coffee shop in Chapel Hill, North Carolina, where he was working on his dissertation for UNC's school of information sciences (that's what library schools renamed themselves when everyone thought libraries would soon be extinct). The shop "had really good coffee and no Internet access," he recalls. "You could get much more focused on your work because you didn't have any distractions," and yet it also had "a very social character."

When a nearby store set up a free wireless access point, though, things started to change. As word spread about the wi-fi, more people would "sit at their laptops for hours on end." The atmosphere became chillier and more businesslike. And between that and fingertip distraction, it was harder to work.

Stutzman needed a way to keep focused while writing his dissertation—perhaps ironically, he was researching how people used social technologies when transitioning from high school to college—and he came up with the idea of a program to disable his computer's Internet access. On a Mac, it's easy to turn off the wi-fi card. The problem is that it's just as easy to turn it back on. He needed something that even a programmer like himself would have trouble getting around.

It took "maybe two hours" for Stutzman to hack together the first version of Freedom. After using it himself for a while, he decided to put it online. A few thousand downloads later, it was clear that Freedom had struck a nerve. Reporters started calling. Writers like Nora Ephron praised it for improving productivity. Naomi Klein, the activist and author of *No Logo* and *The Shock Doctrine,* tweeted its praises (ironically).

Programs like Freedom are usually discussed in the context of life-hacking or other movements to improve productivity. In some measure, they're an information-age update of the scientific-management movement led by Frederick W. Taylor (Taylorism is named for him) and Lillian and Frank Gilbreth (who are best remembered from their son's book *Cheaper by the Dozen*). At the turn of the century, Taylor, the Gilbreths, and their students aimed to make workers more efficient by conducting time-and-motion studies of manual labor, redesigning factory work flows, carefully planning and regulating the work of laborers, and implementing programs that financially rewarded highly productive workers. Taylor envisioned factories and supply chains that worked with machinelike efficiency; to realize that, as he famously put it, "the system must be first." Lifehacking, by contrast, is self-directed and largely service-oriented; while Taylorists gave managers tools for optimizing the efficiency of manual workers, assumed that workers were fundamentally shiftless, and sought to translate faster work into higher corporate profits, lifehackers want nothing more or less than to redesign themselves for their own benefit.

Fred Stutzman released the first version of Freedom in 2008, and word of mouth attracted thousands of people. Two years later, he released a commercial version, but the program really wasn't changed much. Adding more features to it wouldn't make it more useful, he thought, just more complicated, and that's not what most users wanted. Likewise, LeechBlock's underlying code has improved over the years, but the latest version looks a lot like the first. When Herraiz and Soto upgraded OmmWriter, the team ignored calls for new features and "decided to keep focusing on improving the audio and visual experiences," commissioning new pictures and music rather than adding more fonts or tools. They've all realized that Zenware works because of its limits, and ambitious upgrades would be counterproductive. The Zen and minimalist aesthetic reminds users to stay on track.

Stutzman has several years' feedback from paying users of Freedom, and the reasons he's found for their using it helps us understand the virtues of Zenware as a whole. First, Freedom isn't complicated. Things like privacy and productivity are "computationally hard problems," he says; they're "far too complex to understand completely," the way you can understand conventional computer science problems by breaking them down into smaller tasks, routines, and algorithms. Humans just work on too many different things, in too many different ways, for any single piece of software to work for everyone, all the time. "We've got lots of complex systems and procedures to manage productivity," he notes, but none of them work well for everyone, and lots of them are designed to improve organizational productivity by enforcing common work patterns. Creative work is *intractable,* a fancy term for something that can be explained in theory but can't be completely described, decoded, and optimized in practice.

But Freedom doesn't make itself useful by solving every problem faced by every working person; that would lead to an enormously complex system. Nor does it force people to completely change the way they work. Freedom makes itself useful by doing one simple thing and trusting that users will be smart enough to figure out how to use it well for themselves. The program is "not asking people to reinvent their notions of work, or change their practices," Stutzman explains. The program stays very simple, so "people can develop their own system," their own ways of working, "and fit Freedom into it."

The Zen dimension of Zenware also matters. OmmWriter's references to Buddhism and contemplative spaces help users make sense of their experience with software—the way they interact with the software, the way they think about the software, and the way they think about themselves. One fan describes using OmmWriter as being like "writing in a Zen garden." Donald Latumahina, a programmer and teacher in Indonesia, wrote in 2007 that using Zenware "gives me a peace of mind, a 'mind like water.' It is a good condition for me to enter the 'flow' state."

Users also echo the spatial language that Grosjean, Soto, and other Zenware developers use. As Michael Grothaus, an American-born technology writer living in England, describes it, OmmWriter "puts you in the middle of a secluded snowy landscape on a foggy winter's day" where "the words you type appear on your screen as if you were writing them in the sky." He adds, "When I write in it, within minutes I no longer hear the sounds of busy London city life zooming past my flat. It's just you and your thoughts for miles around." Other users echo that sense of being removed from both the regular, distraction-prone world of computing *and* the flow of regular life, instead entering a "creative writing environment." One Catholic priest compares Zenware to the library where he wrote his dissertation. He worked at a wooden table surrounded by books with a "bright light shining on my work space, focusing my attention"; there was "carpet on the floors, and the bookshelves muffled all sound, so my mind could fully concentrate on the task at hand."

But perhaps the most important reason Zenware works is that users *want* it to work.

There are lots of lightweight, inexpensive alternatives to Microsoft Word, and users with a high level of technical proficiency can install Linux and open-source tools, many of which have functionalities that rival commercial word processors'. You don't adopt Zenware unless you want a distraction-free experience: the sense of entering a placid, protective space that reminds you of the sacredness of your own thinking, that won't clutter your screen with unnecessary functionality, that appreciates the value of your attention. (Indeed, one Buddhist nun compared Zenware to an alarm clock. "The alarm clock wakes you up," she explains, "but it's up to you to get up when it does and not just turn it off.") Zenware is partly useful for its formal properties but also because it represents your determination to focus. Further, in the course of acquiring and learning to use OmmWriter or WriteRoom, you're exposed to the Buddhist-like language on the Web site, in the user testimonials, and even in the software itself. This isn't just window dressing; it's what Berkeley anthropologist George Lakoff calls framing—the material signals the developer's intent, sets your expectations, and gives you a language for describing why you use this software.

Freedom is also powerful because "people are making a contract with themselves" when they use it, Stutzman says. Downloading it shows that you take the problem of digital distraction seriously and want to do something about it. Users' sense of commitment actually got stronger after Stutzman started charging for Freedom: "The fact that there's a price helps the contract," he says. Even if it's not dramatically different from the older free version, the paid version is more effective because users take it more seriously. Trying to disable it once it's running also reminds users of that contract, he thinks. Restarting your computer to regain your access necessitates "a moment of reflection: you're forced to sit there and think about why things have failed." Indeed, one of the surprises I found when I started using it was that I didn't just reach for my iPhone or iPad when Freedom was on. *No,* I'd tell myself. *I'm offline for a reason.* For me, the idea of Freedom as a contract with myself was powerful.

The self-awareness these programs promote helps explain why, even years after its invention, Zenware continues to work. Donald Latumahina said that four years after discovering JDarkRoom, "having a distraction-free environment still makes it easier to focus." James Anderson finds that LeechBlock still works for him. "I know how to get around it, of course," he says, "but it still slows me down enough to serve as a disincentive to bunk off. I also think it's had a kind of training effect on me." Fred Stutzman continues to rely on Freedom; he knows how to get around its restrictions, but he doesn't try to. Michael Grothaus still uses OmmWriter and likes "the feeling of being in a snowy field miles away from anyone" that the program provides. (Though he adds that at this point, for more introspective writing, he's "pretty much gone back to keeping an old fashioned Moleskine journal," which he finds "superior to anything the computer has to offer as far as distraction-free writing goes.")

Users describe Zenware as inspiring a "mind like a mirror"; it is not something that they *consume* but an experience that they help *create.* "People want to believe that this will work," Stutzman explains. Their desires and expectations, as much as the software's functionality and UI, help make Zenware powerful.

This suggests an important point. Building up your extended mind isn't just about adding newer, more sophisticated technologies or offloading tasks into the Cloud. It's about choosing and using technologies that help you build habits of mind and cognitive abilities that externalize and thus reinforce mental capabilities. Just as a seasoned pilot's judgment and experience can enable him to get a plane through problems that automated systems can't (or that automated systems cause), so too the mental skills we develop are likely to be more flexible and adaptable than their digital equivalents.

And the success of technology turns out to depend on willing participation. Landscape architect and multimedia designer Rebecca Krinke has spent years exploring architecture and new media that encourage people to be more contemplative. "How," she wants to know, "do we interact with devices in a way that doesn't drive us crazy?" She thinks that while landscape architecture can show us how to create technologies that help users be calmer and more thoughtful, "you have a responsibility to do it yourself." After a decade of working on this problem, she's concluded that there's no perfect technological fix. The transaction between the person and the technology, not the technology itself, is what's key. In an important sense, contemplative spaces are verbs, not nouns: you can design a place that's meant to support contemplation (and we'll learn what that is and how to use it later), but it comes alive only when people are able to use it to calm their minds. A Zen garden isn't a Zen garden if no one's in it.

This means if you're going to practice contemplative computing, you need to know how to be contemplative.

THREE

MEDITATE

Sit on a cushion, cross your legs, close your eyes, and take a deep breath. Let it out slowly. As you do, relax your mind. Draw a breath in, using the muscles in your stomach to expand your lungs, and slowly count: *One... two... three... four.* Try to think about only the numbers and your breathing, nothing else. Hold it for another four. Then exhale, counting again. Take another breath, stay focused, and count again. At some point, your concentration will fail, and your mind will wander to something other than your breath and counting. It fails for everyone eventually. Don't be discouraged. Let the distraction pass, recenter yourself, take another breath, and start over.

This is a simple version of vipassana breathing meditation. It's meant to serve as an introduction to the practice and to help you see the difficulties and virtues of meditation. Meditation of many varieties is practiced all over the world. The popular image of meditation is that it's a kind of blissed-out blank-mindedness. Nothing could be farther from the truth. I've done it for years, and every session is a challenge. It's taught me a great deal about how my mind works. It's essential to my mental health and to developing the tools necessary to practice contemplative computing. I get a lot of benefit from it. Even though I'm actually a terrible meditator.

In the predawn hours, before anyone else in the house is awake, I put on my noise-canceling headphones and sit in the living room. Part of me feels the same sense of calm, pleasant excitement I have while waiting to board a flight. A full lotus position is beyond me; sitting cross-legged is the best I can muster. (Fortunately this is also the Burmese position, so I can hide my weak practice behind an exotic-sounding name.) Once I'm settled, I turn on the Insight meditation timer on my iPhone, close my eyes, breathe deeply, and for the next hour work to make my mind like a mirror.

What is it that we seek when we meditate? In our everyday lives, activities that absorb our attention, focus our minds, and bring a sense of calm and purpose are described as contemplative. Driving, cooking, listening to music, skiing, caring for the sick, swimming, praying, sitting by a river—virtually any activity can be an opportunity for real-world application of contemplation and mindfulness. But what's especially useful about meditation is that it isolates the phenomenon of contemplation, allows us to explore it in depth, and advances the ability to be contemplative. It's practice, study, and self-observation all at once.

My body spends the first few minutes settling. It's not like sleeping; my body relaxes but it is poised, not limp. It takes energy to sit properly (or even improperly). After I'm centered physically, it's easier to clear my mind and begin the meditation in earnest. Contemplation has an embodied dimension: it seems to be all in the mind, but, like any cognitive process, the body supports it.

I sit still and breathe slowly, drawing in air over the course of several heartbeats, holding it for several more, then letting it out. But my mind doesn't want to settle down and be calm. Like a child who doesn't want to go to sleep, it throws out images and memories, struggling to keep itself active. I let the tantrum pass, resettle, and move again to calm the surface of my mind. Sometimes it works, but most of the time it doesn't.

There's nothing quite like meditating to show just how active and random the everyday mind is. When people are bored, they think the problem is that their minds are unoccupied, and often people shy away from being alone with their thoughts. But when you sit and actually *try* to clear your mind, you realize that even at your most bored, you hear a silent monologue, a kind of cognitive channel-surfing. It's very hard to turn it off. As I sit, focusing on settling my mind, my mind throws out bits of an episode of *Lost,* the cover of Led Zeppelin's *Houses of the Holy,* the amount of money I should send to the credit-card company, an e-mail from someone I'm interviewing, a scene from *The Hunger Games* (that one is my daughter's fault), a blog post I wrote years ago about the Nunberg error (named for Berkeley-based information scientist Geoffrey Nunberg), the cover of Susan Blackmore's *Zen and the Art of Consciousness,* a photo I took of the window of a bookstore in the cathedral town of Ely.

The timer chimes, signaling that five minutes have passed. My focus swings to the bell, and I concentrate on the fading sound to the exclusion of everything else, and as it vanishes, I imagine I can still hear it in my mind. A minute later I realize I've been thinking about snow cones and my Netflix queue. I fail constantly. I let it go. Then I start over again.

Eventually, though, my mind begins to quiet. I can feel everything slow down. I lose track of the chimes. I don't know how many are left, and I don't care. I focus now on a very modern kind of image: a picture of my own brain, like an fMRI, with thoughts flashing across it in angry red. As my mind slows, the red fades, and as my concentration increases, my brain begins to glow faintly white. Another unbidden thought; another trace of red that recedes like an afterimage. If it goes really well, the glow continues, and I feel the sort of exhilaration that comes when hard effort is paying off—when you reach the end of the steep trail, stand at a peak, and can see miles in every direction. But some part of me is careful not to enjoy it too much or too consciously. If I focus on it, it disappears. To sustain it, I have to just be present with it.

Whether contemplative activities are practiced on a mountaintop or in a studio, car, or kitchen, they share a critical property: they take you into a state of detached, calm engagement. It's the calm that requires you to draw on skill and self-control. It's not passive; it's active. My body is never completely relaxed when I meditate, nor is it passive; sitting up, controlling my breathing, centering my mind, and concentrating all require energy. The closest physical state I can compare it to is that moment in a karate competition before the action starts or the whistle blows, when you're poised and you feel the energy flowing through you, and, without really thinking about it, you know you're ready to strike or block an opponent's move.

For me, meditation is hard; it's much more like working out at the gym than spacing out. I have a difficult time reaching that state of complete serenity that I imagine more experienced meditators can sustain for hours.

For all the poor quality of my practice, though, I still benefit from it. Learning to restart something without prejudice is an immensely valuable talent. One of the most important skills meditators must develop is the capacity to continue practicing in the face of constant failure. I get a lot of practice at this, bringing my wandering mind back from random thoughts. In the classic *Zen in the Art of Archery,* Eugen Herrigel's teacher urged him always to take his next shot unburdened by previous failures to hit the target; as he improved, his teacher urged him not to be influenced by his successes either, to stay in the present moment. There are lots of times that concentration fails, a diet falters, or a project needs a reset; meditation allows you to practice starting over.

If I can clear my mind entirely for even a heartbeat and think of nothing for just a second or two, I can then think about one thing for a very long time, and my mind can spend hours wrapping itself around a challenging problem. I reach a stage where I feel my mind working through problems, turning things over, without my conscious involvement. My calm mind is able to do things that my ordinary mind never can. It can concentrate relentlessly. When I'm deep in this state, I never completely disengage from whatever problem I'm working on; even when I'm shopping or doing the dishes, I can feel a part of my mind going over the issue.

And the desire to feed on distractions—to check back with Facebook, to see if anyone has reposted my last tweet, to see if Paul Krugman has said anything new on his blog—recedes. The feeling I have as I sink into the heart of a session, when I can sense my mind shifting gears, stays with me.

When it goes well, meditation seems to be a kind of intense flow state. Its goals are simple, yet achieving them is inexhaustibly challenging. It stretches time in ways that are strange and wonderful. It's hard but also immensely pleasurable. It's a purer version of what happens when I'm working well: focus is intense and effortless, my whole mind—both my conscious mind and the more mysterious part that generates interesting ideas and elegant turns of phrase—is tuned to the problem, and I can feel a solution within reach. I asked Mihaly Csikszentmihalyi, the father of the concept of flow, about the connection between the two. "It's true that much flow does involve a reflective, meditative stance," he says. "Meditation can be a form of flow, and flow can be a form of meditation." But, he continues, in flow, you engage with things in the world—chess pieces, salmon, bows and arrows, motorcycle repairs; in meditation, "both the challenge and the skill are inside of you. That makes it very hard, because you have to master your own inborn need for novelty and movement and the monkey mind."

True. But sitting quietly, watching my thoughts, I feel as if that internal world isn't completely unitary. In our everyday lives, we talk about the mind and the self as more or less unbroken wholes. But when I'm meditating, it feels almost as if there are different parts to my mind: one controlled and directing, the other a random generator of chatter. Tibetan Buddhism contends that there's no unitary mind, no *self,* but eight parts that work together to generate the illusion of a stable, enduring selfhood. The senses constitute the first five parts. Analytical, logical capabilities make up a sixth. The monkey mind is a seventh. The eighth part is self-aware, focused, and can control the other seven. The challenge in meditation is to strengthen the last and tame the seventh. Monkeys don't like being held still, even when they're in your mind.

Meditation is the original neuroscience, the world's oldest conscious exploitation of neuroplasticity, and it's a twenty-five-hundred-year-old answer to the twenty-five-year-old problem of digital distraction. It helps you restore cognitive abilities that electronic overindulgence can erode. It shows that you can change your extended mind from the *inside* (through contemplative practices) as well as from the *outside* (through more careful technology choices).

The fact that meditation should be useful in dealing with digital distraction won't come as a surprise to the millions of people who meditate regularly or to scientists who've studied the therapeutic benefits of meditation. In the 1970s and 1980s, psychologists began to apply contemplative practices to therapy, most famously by developing mindfulness-based stress reduction (MBSR), a technique that uses meditation to combat chronic stress. Since then, contemplative practices have been adopted in fields that require high degrees of creativity and concentration and an ability to perform under pressure. Educators are integrating contemplative practices into fields as diverse as science and jazz. Coaches use meditation and visualization techniques to sharpen the performance of elite athletes. Military trainers and psychologists use contemplative practices to improve combat performance and to ease posttraumatic stress disorder. Organizations use contemplative practices to improve collaboration and communication and to smooth dispute resolution. Even lawyers use contemplative practices to improve negotiation skills and invest the practice of law with spiritual meaning.

The social and psychological benefits of meditation are well documented. For a long time, though, it was very hard to say what actually happens when someone meditates. Its intense subjectivity made it impervious to science—that is, until tools like the EEG and fMRI let us see what was going on in meditators' brains and allowed researchers to connect the subjective experience of meditation with objective observations of activity in the brain.

What we're learning is that meditation doesn't just make the brain behave differently temporarily. It rewires it.

Much of the pioneering work on the neurological effects of meditation was done at a University of Wisconsin lab led by neuroscience professor Richard Davidson. As a graduate student at Harvard, Davidson had spent time with Ram Dass, a former Harvard psychologist and a collaborator of Timothy Leary's. Davidson took time off from graduate school to study meditation in India. In 1992, the Dalai Lama encouraged Davidson to undertake neuroscientific studies of monks; he didn't have to suggest it twice.

Davidson started looking at what happened to monks' brains when they meditated and whether meditation could produce long-term structural changes in the brain. It was very much an open question whether he and his fellow researchers would detect any neurological changes in monks' brains; the concept of neuroplasticity—the notion that structures in the adult brain can change as new tasks are learned and expertise acquired—was still fairly new.

Davidson's team also wanted to study a phenomenon called gamma synchrony. First observed in EEGs in the early 1960s, gamma waves are neural oscillations that appear to sweep across the brain. They're particularly noticeable when working memory and perception are being used and during periods of intense concentration; scientists have observed heightened gamma synchrony (that is, many gamma waves occurring at the same amplitude and frequency) in rats who are exploring mazes, in rhesus monkeys watching computer screens, and in musicians listening to music. Depending on their strength, the gamma waves can act on particular regions of the brain—the visual centers when someone is working on a puzzle, for example—or on the entire brain. By providing a kind of standard time that all different parts of the brain can work from, gamma synchrony may help the brain construct a unified experience of reality from diverse sensory inputs—in other words, it may serve as a foundation for the emergence of consciousness.

In their first study, Davidson and colleague Antoine Lutz connected monks (and a control group of college students) to EEG monitors and monitored their brain activity as they went through a series of meditation exercises. EEG (short for electroencephalogram) uses sensors attached to the scalp to measure electrical activity in different regions of the brain. One of their first subjects was Matthieu Ricard, a biochemist turned Buddhist monk who was himself an expert on the scientific study of happiness. When Davidson instructed Ricard to meditate on unconditional love and kindness (one of several types of meditation practiced by Tibetan Buddhists), the EEG recorded large increases in gamma wave activity and pronounced increases in activity in a section of the left frontal lobe that Davidson had previously identified as generating compassion. The increases were so large, in fact, that the scientists assumed their equipment was malfunctioning. But when they repeated the experiment, it became clear that this was how the monks' brains worked; after years of intense practice, the monks' brains operated in a well-coordinated manner during meditation, producing patterns that corresponded to strong attention and memory.

Davidson's team reported its findings in the prestigious *Proceedings of the National Academy of Sciences* in 2004. Davidson and his group have continued studying monks, and they've conducted other studies on people who are new to meditation or who use contemplative practices to deal with psychological or medical problems. They've established that meditation can positively—and permanently—affect brain function. Meditation, like playing the piano or violin, strengthens parts of the brain, just as exercise strengthens certain muscle groups and reflexes. In one sense, these results aren't surprising: changes in brain function have been observed in mathematicians, jugglers, musicians, and London cabdrivers (who need outstanding visual memories to navigate the city's streets). "If you do something for twenty years, eight hours a day, there's going to be *something* in your brain that's different," neuroscientist Stephen Kosslyn points out, though he admits he's "amazed" at some of the things monks seem able to do.

While Davidson and his colleagues were covering monks' heads in EEG sensors in Madison, Wisconsin, neuroscientist Clifford Saron and his colleagues were on a mountaintop two hours north of Denver, Colorado, working in a lab underneath the main hall of the Shambhala Mountain Center. Saron is a professor at UC-Davis and the director of the Shamatha Project (*shamatha* means "calm abiding" in Sanskrit), one of the longest-running scientific studies of meditation. On the floor above Saron's group, thirty students were going through an intensive, three-month course in meditation with Alan Wallace, a meditation teacher whose life's path has taken him from a childhood in Southern California, to a spiritual pilgrimage to Dharamsala in the late 1960s, to years as a Tibetan Buddhist monk, to a PhD in religious studies from Stanford, and finally back to Southern California, where he runs the Santa Barbara Institute for Consciousness Studies.

Davidson's lab has made some spectacular discoveries by studying monks who've spent tens of thousands of hours meditating. The Shamatha Project uses many of the same tools—a mix of EEG and psychological tests, most notably—but rather than working with expert meditators, the project focuses on a population of sixty students from their first days at Shambhala. The idea is to use the student's own beginner's mind as a baseline to measure the impact of meditation and as a way for the group to better understand what happens in the brains of novices. After students leave the center, some continue meditating regularly, while others drift away from the practice. All are retested at regular intervals via laptops loaded with experiments and sent through the mail. (The project spends a small fortune on postage.)

What Saron and his colleagues want to measure is the long-term effects meditation has on concentration, attitude, and health. The Shamatha Project is halfway through its ten-year life, but it's already generating some interesting results. On perception and attention tests, subjects demonstrate increased ability to resist distractions (what psychologists call response inhibition) and increased ability to focus and maintain their attention during the kinds of dull experiments that scientists like to design. They also report having greater self-control and feeling more adaptable. These results confirm earlier experiments and reports of clinicians, but by working with a larger population for a longer period, the Shamatha Project can more accurately measure how long meditation's benefits last.

Even more surprising are the results from the blood samples, which were taken at regular intervals so scientists could measure telomeric length. Telomeres are DNA sequences found at the ends of chromosomes; they're a bit like the plastic ends of shoelaces in that they prevent chromosomes from fraying or becoming damaged. Every time a cell divides, the information-bearing sections of chromosomes are copied exactly, but the telomeres get a little shorter; when they become too short, the cells stop dividing. Scientists believe that telomeric shortening contributes to aging, and that slowing the shortening process could extend human lifespans. And that's what's striking in these results—Shamatha Project alumni generate more telomerase, an enzyme that builds up telomeres. In other words, at the cellular level, they seem to be aging more slowly.

This isn't the first time scientists studying meditators have observed health benefits. Participants in one eight-week meditation study had improved immune responses to influenza after the end of the experiment. This study is promising not just because it offers hope for people who hate needles. Very few people get to spend weeks under the tutelage of world-class meditation teachers, and even fewer can become monks, but the immune study shows that modest but tangible changes can come from only a few weeks of half-hour sessions. Studies of people who've gone through mindfulness-based stress reduction programs have shown that participants exhibit more activity in the left anterior sections of their brains after eight weeks of meditation and have more positive moods. Working memory also improves, maybe because mindfulness involves paying constant but casual attention to one's own mind, and *that* requires observing and remembering fleeting information about the mind's prior states. (Try this experiment: Remember the most recent thing that distracted you. Since it was probably just a moment or two ago, you might expect that to be easy. Is it?) Meditators expend less effort maintaining attention than people who do not meditate. At the same time, meditators' attention does not get as stuck on a single stimulus. It may also improve basic perceptual ability, leaving the mind more energy to concentrate on other things.

In other words, the subjective benefits of meditation are accompanied by physiological changes in the brain. Those changes improve other cognitive functions, like memory and attention, and strengthen emotional balance. And these changes are lasting, not temporary. But rather than your just trusting that these techniques can help you deal with the distractions and frustrations of online life, let's look at a group of people who are regular users of social media but who seem immune to its effects. They spend hours a day online without the media feeding the monkey mind. They maintain relationships with information technologies that keep them firmly in control. And they have a perspective on digital distraction that is unique.

Meet monks who blog. They're ordained Buddhist monks who spend hours a day in study and meditation and hours more posting instructional videos on YouTube, writing blog posts, managing discussion groups, and using Facebook and Twitter to share devotions and lessons. Some are members of orders that let them pursue secular lives and families, some live alone in the jungles of Asia, and some live in monasteries. All are pursuing a rigorous, ancient discipline that promises an escape from desires, distractions, and worries. Yet they're as familiar with smartphones and social media as they are with the Four Noble Truths.

Every religion uses the Internet to evangelize among nonbelievers; instruct the faithful; carry on sectarian debates; and manage the everyday work of organizing worship, charity, retreats, pilgrimages, teaching, and study. Buddhism is no different. The religion has about 350 million practitioners worldwide. In some countries, such as Thailand and Japan, it's central to the national culture and identity, and monks today offer prayers in temples that have been in continuous use for over a thousand years. While Buddhism has deep national roots, in the twentieth century, the *sangha* (which means "the community of the Buddhist faithful") has become more mobile and cosmopolitan. Some of these changes were the results of war and revolution. Some monasteries in Tibet, Vietnam, and Cambodia were closed by Communist regimes during the Cold War. Monks (most famously the Dalai Lama) joined the vast population of émigrés and displaced persons relocating to Europe, Australia, or North America. Entire Tibetan schools were refounded in India; in the past fifty years, the communities of Dharamsala and Namdroling have become vital global centers of Buddhism. (Imagine if the British had lost World War II and refugee Oxford and Cambridge dons had to refound their colleges in the Canadian Rockies—you get the idea.) Exchanges between monastic orders have encouraged cross-fertilization between formerly separate schools of Buddhism, and interfaith dialogues and collaborations between monks and scientists have raised awareness of Buddhism in Western circles.

While the popular perception of Buddhism is all saffron robes and incense, the religion has adopted and used information technologies for millennia. Buddhist monasteries experimented with block printing and xylography (carving or engraving blocks that mix images and text) from the 600s. The world's first printed book was the Buddhist Diamond Sutra. Chinese monks and scholars used the technology in decades-long projects to print the Tripitaka and sutras, and they carried the book to Turkestan, Mongolia, Japan, and Korea during the tenth century CE. Given their deep historical involvement in printing technology and their contemporary needs as a globally dispersed community, Buddhists, unsurprisingly, appreciate the Internet's value for communication and coordination.

Buddhist social-media users, bloggers, and Web site creators had simple reasons for developing an online presence. As Yuttadhammo, a blogger monk, explains to me, "If you want to share something, you have to go where the people are." After his first experimental, bare-bones YouTube video attracted a thousand viewers in a week, he realized that the medium could be a powerful way to reach followers. Other monk bloggers are part of the first generation of digital natives.

They learned about meditation and Buddhist teachings through Web sites and discussion groups. One American-born blogging nun discovered the Zen Mountain Monastery, where she was later ordained and lived for eight years, through its Web site. So central is the Internet as a resource today, Yuttadhammo says, that "writing books on the dhamma is a bit pointless these days unless you offer a PDF file as well."

Monk bloggers see the Web as an amazingly valuable publishing tool. "Instant access to the Buddha's teaching has been a great boon" for novices and monks alike, one says. Another monk tells me that the Internet "played a huge role in my finding teaching, contemplating it, memorizing it, and relaying it." An elderly monk who uses a Kindle describes its lightness and portability as a great blessing.

The utility of the Internet for building virtual communities and strengthening existing *sanghas* has also driven experimentation with the Web. "Having a community is an absolutely foundational part of being a Buddhist practitioner," Lauren Silver, a Buddhist and educator at the Computer History Museum in Mountain View, California, explains. While there certainly are monks who live largely in isolation, she continues, "Buddhism was founded and has thrived on the integration of the teachings and practices into communities." Many meditation centers and temples have Web sites, and the most active sites have a global reach. When Lauren answered e-mail for her local meditation center, she recalls, "What was always amazing was that we would get messages from people who live in Latvia or the Australian outback, saying 'I never ever met a Buddhist, I've been meditating ever since I found these teachings, and I have this one tiny practice question.' " So powerful is the Web's potential for enhancing communities that the challenges of distraction are worth meeting.

At the same time, monks are skeptical that online communities can ever rival real ones and doubt that virtual experiences can be as rewarding as the hard daily work of following the Noble Path. Buddhists place great store in "the practice": they talk about it much the same way elite musicians talk about practicing their instruments, an inescapable foundation needed for all high-level work. The Buddha challenged followers to test the value of his teachings for themselves, not to take them on faith. As Yuttadhammo puts it, "Buddhism is a path inside, not an outward expression. The Internet is a resource, not a part of one's practice." One practitioner I interviewed gave away his computer when he felt he had read enough and needed to spend more time meditating. "I was finished with it," he said simply.

"Words will never have the power of an actual experience," an American Buddhist nun tells me. And a Finnish Zen monk argues that while the "virtual world can help Buddhists in their practice, it's nothing similar to the practice in real life." As Yuttadhammo says, "We should never use the Web as our sole source of dhamma practice; I think it would be delusional for us to think that our online dhamma community is really the most important aspect of our individual lives as Buddhist meditators." As a result, the Web is seen as a gateway to deeper practice, not an end in itself.

Monks' and nuns' personal time online tends to be tightly bounded. The rhythms of monastic life put limits on when they're online. "I'm very busy and simply have other priorities; I can't afford to sit and watch cute cats indefinitely," one says. Several monks and nuns I talk to go online only in the evening. Most of them have desktop computers rather than laptops, which makes it easier to separate their real-world and digital lives; the few who have laptops keep them on their desks or, in one case, a closet. They use older machines, which reflects both their more utilitarian views of technology and their tighter budgets.

Cell phones are decidedly unpopular. Choekyi Libby, the first native-born Israeli Buddhist nun, says, "I have a cell phone, but I'm not glued to it. I don't use it very often. I'm not constantly reachable. I like not being connected all the time." Bhikkhu Samahita, a Sinhalese monk, was once given a cell phone by a follower, but it was a curiosity to him—*Who would I need to speak to?* he wondered—and besides, the reception at his hermitage was terrible. Yuttadhammo uses the camera on his phone, but Sri Lankan carriers don't tend to place cell-phone towers around monasteries, so reception is weak. Elsewhere, monastics have cell phones instead of landlines, but like the computers, the phones stay in the monks' rooms. Phantom cell-phone syndrome is virtually unheard-of in this community.

Monks see being online as an opportunity to practice Buddhist precepts. Damchoe Wangmo says that keeping up with Western news is "a good basis for generating love and compassion," as it "makes my own problems seem small in comparison." Yuttadhammo sees online interaction as quintessentially Buddhist. It allows him to do good without attachment. "I don't really make connections with other people online; I just help the person or people I'm interacting with and am not troubled by letting go and moving on once that is done." I'm reminded of the story of two monks who find a beautiful woman standing at a river crossing. The older monk lifts her up and carries her to the other side. The younger monk is outraged and, after fuming silently for hours, asks how the older monk could violate the rule forbidding them to touch a woman. "I put her down miles ago," the older monk replies. "Why are *you* still carrying her?"

Viewing online activities as important but secondary to real-world practice, taking a utilitarian view of information technology, and using devices in ways that minimize entanglement help monk bloggers stay in charge of their technologies and keep them from feeling overwhelmed by their devices. But how do they manage blogs, tweet, answer questions from novices, e-mail fellow monks, conduct online meditation sessions, and troubleshoot erratic Internet connections without losing their sense of inner balance or their capacity to focus?

Consider Bhikkhu Samahita, whose posts on Buddhism ricochet through the social-media world: news of a new essay—he writes one every day or two—appears in minutes on Facebook walls, Twitter feeds, Google+ pages, and online discussion groups. Even political campaigns have a hard time messaging as quickly as this. Samahita (*bhikkhu* is the Pali word for "monk") is the creator of the influential Web site What Buddha Said; it receives tens of thousands of hits a year, with visitors from the United States to India to Malaysia. Eight thousand followers worldwide receive his daily devotions, consisting mainly of glosses on the Pali Canon, the oldest and most revered texts in Buddhism.

To keep all this going, Samahita spends several hours a day online, in the early morning and late afternoon. This is an impressive whirl of activity, but it becomes even more mind-blowing when you realize that he does it all from a small hermitage on the island of Sri Lanka. Samahita is a forest monk, one of several thousand Sinhalese monks who model their lives on the Buddha's years in the forest in search of nirvana ("awakening"). Forest monks live in huts, caves, or small concrete houses. By tradition, their homes are at least five hundred bow-lengths from a village, out of earshot of village people and out of sight of other dwellings. Like the early Christian desert fathers, forest monks seek an especially pure, ascetic way of life; none of the luxuries of monasteries, towns, or other people for them. They sleep four hours a night, meditate for eight, and follow the 227 rules that guide every aspect of monastic life.

For the past decade, Samahita has lived at Cypress Hermitage, a small whitewashed house forty-two hundred feet above sea level in the mountains of central Sri Lanka. You reach Cypress Hermitage by driving through a tea plantation and then getting out and following a dirt track too small and steep for cars. The map to the hermitage carries a warning to visitors: "Stay inside Tea Bush Plantation. Don't go into forest. Go only up. Never down." Samahita sees other people maybe one day a month, when he walks into town for supplies; visitors come to the hermitage once or twice a year. Yet he spends four or five hours a day in front of his HP Pavilion dv7 laptop. His computer is flanked by large windows, and the views can be spectacular. The laptop and Internet connection are powered by a solar-panel array and a microhydroelectric generator that draws on a nearby stream. (Think about that the next time you feel tech savvy and ruggedly independent because you changed your own printer cartridge.)

How does he maintain these two seemingly very different lives? We began a conversation by e-mail. What's rewarding, and what's challenging, about being a forest monk? I ask. "Completing the Noble Way is both the most rewarding and challenging!" he writes back. What is life in the forest of Sri Lanka like? "Peaceful. Calming. Happy. Simple," he replies. "Smiling is the forest." Other answers are a mix of telegraphic prose, quotations from poems, and hyperlinks. His English is impeccable, but I get the distinct impression I'm interviewing someone who no longer has much use for words. It's a bit like talking to Yoda. A tall, Nordic Yoda.

Samahita was born in Denmark, and before his ordination, he was a physician, a specialist in tropical and infectious diseases. He was also a professor of bioinformatics at the Technical University of Denmark. Researchers in bioinformatics develop tools to analyze vast data sets related to medicine and health; they look at everything from strands of DNA to World Health Organization statistics on emerging epidemics to Walmart's sales of cold and flu medicine. For an ambitious researcher at the turn of the millennium, it was a great field to be in. But he was unhappy with his life and began to have problems with depression. Despite being a doctor, he resisted taking antidepressants. A chance encounter with a Tibetan monk led him to try meditation. It cured his depression and gave him a glimpse of a new life. He started What Buddha Said in 2000. The next year he exchanged a laboratory in Copenhagen for a monastery in Sri Lanka. After two years' study, he was ordained, and he moved to Cypress Hermitage.

It's a remarkable transformation, and I take it as a promising sign. If someone accustomed to wrangling terabytes of data and being on call at a hospital can go from a life of technology-driven distraction to one that balances the quiet of the forest with a life online, then maybe the rest of us can learn to be a little more contemplative in our use of technology.

I ask if there is a paradox in using the Web, a medium so many people find to be a great distraction, to teach Buddhism, which concerns itself with eliminating distraction and desires. "If one does not crave it and utilize it rightly (not so easy) then the lotus can grow even in the mud," he replies. The lotus is a symbol of purity in Buddhism because its beauty can flourish even in dank areas, and its petals have a remarkable ability to resist dirt (thanks to a unique nanostructure that scientists have only recently modeled).

Of course, plenty of people would describe themselves as addicted to their devices—the nickname "CrackBerry" didn't come out of nowhere—but in Buddhism, craving (or *tanha,* which literally means "thirst") is at the root of suffering. As the Buddha put it, craving "is bound up with impassioned appetite," and it "seeks fresh pleasure now here and now there." Feeding such desires temporarily sates them, but they return with a vengeance, hungrier than ever. Not a bad description of the woman who says, "I'll just check my news feed and hop off," only to find herself still there an hour later.

So, I ask Bhikkhu Samahita, can a person who spends four or five hours a day online really *never* mindlessly surf the Web? He doesn't seem to understand the question at first. "Whether internal (memories, flashbacks) or external (world, IT, TV), it has to be dealt with accordingly."

All distractions are the same, he says. It doesn't matter where they come from.

I suspect I'm not making myself clear, so I try again. Does the Internet pose special challenges? "The beauty and peace here make the Internet dull and noisy in comparison," he says.

I have friends who can hardly make it through a traffic light without checking their e-mail. Yet here is an MD, a former professor from one of the best universities in Europe—in short, exactly the kind of information-saturated alpha type who you imagine would have trouble being out of touch for even a minute—serenely declaring that a two-room whitewashed house in the middle of nowhere makes the Internet seem dull.

Clearly, something interesting is going on here.

I interviewed another monk. Yuttadhammo is a monastic social-media entrepreneur. His "Truth Is Within" YouTube channel has had over a million visitors. His videos range from instructions on meditation to discussions of Buddhist scriptures to tours of an unfinished retreat center near his hermitage. He also answers questions that viewers send him: Should one kill pests? asks one, and How do monks get their Buddhist names? asks another.

The videos are shot on a Canon Vixia HF200 that was donated by an American student of Yuttadhammo's. Most monks' technological devices are donated by followers, either directly, as gifts, or through contributions to charities set up to support them. The "Ask a Monk" videos are one element of a portfolio of Web sites, wikis, Internet radio streams, and online study groups that keep Yuttadhammo engaged in a digital conversation with students and fellow monks around the world. "Even some of my meditation is done in an online group," he says. These online activities are now his main work as a monastic. "I don't build huts, but I have built an online community."

Does recording videos, writing blog posts, meeting students online, and answering e-mail ever become overwhelming? No, he says. "There were times I found myself distracted; I think because of how depressing life was at those times. But now that I'm in a comfortable place where I can do my work in peace, I don't have much interest in anything besides a cursory glance at the news every morning."

The fact that Yuttadhammo connects distraction and depression isn't surprising; one of the symptoms of clinical depression is an inability to concentrate. From my own experience, I know that, particularly for well-educated or high-achieving people, depression and distraction can reinforce each other. Depression makes it hard to work, which feeds what Winston Churchill called the "black dog" of mental lethargy. For most of us, Yuttadhammo's "comfortable place" would seem far from perfect; mosquitoes are a constant problem, there are leeches during the rainy season, and snakes and scorpions are a danger. Monkeys—the real ones, not the mental ones—"can be a nuisance," because they're smart and unafraid of people. But these challenges are "nothing compared to the stress of living amongst human beings."

Yuttadhammo lives in a *kuli,* a small hut, a few miles away from Bhikkhu Samahita. Like Samahita, Yuttadhammo is a monk who left a wired life in the West for the rigors of life in the forest. He was born and raised in a "nominally Jewish" household in Canada, and he discovered Buddhism while traveling in Thailand during college. After a year in a monastery in Canada, he returned to Thailand, was ordained, and then moved to Sri Lanka.

Yuttadhammo's immersion in technology puts him at odds with monks who consider it inappropriate for a forest monk to spend so much time online. While monks are supposed to renounce worldly goods, materials for teaching and study are allowed, so one can make a case for having a computer; still, Yuttadhammo admits, "I suppose I am not a true forest monk these days. I'm not really sure what kind of a monk I am." But technology lets him strike a balance between the demands of monastic life, and the desire to teach. "Being online puts me at arm's length away from the world; I can do good in the world and still ignore all of the worldliness." This also explains why he uses YouTube and streaming video but opts out of other social media. "I tried Facebook and couldn't see the point of an antisocial monk acquiring 'friends.' Twitter seems equally pointless."

Five hundred miles northwest of Samahita's and Yuttadhammo's quiet hermitages, in southwest India, stands the sprawling Namdroling Monastery. Damchoe Wangmo is one of five thousand students there; she is in her ninth year of a ten-year program, and after she graduates, she expects to teach and translate Buddhist scripture into English. The monastic college's electricity is available only episodically, and the frequent outages as well as Damchoe Wangmo's studies make having a regular online presence difficult for her. But her blog Nun Sense is written to give Buddhists who are considering ordination "an idea of what monastic life is like." She also runs an online group for her fellow monastics. (That's right: Buddhist monastics have their own darknet.)

Damchoe Wangmo grew up in Canada. Her father was a Presbyterian minister, and her mother taught Sunday school. "I used to finish off the grape juice left over in the tiny communion cups" in their church, she says. In high school she stopped going to church—she didn't believe what she was being taught, she explained to her mother—but she remained interested in religion and soon discovered Buddhism. After graduating from high school, she studied Buddhism in Vancouver, taught in Dharamsala and Taiwan, returned home, pondered her future, and then, in 2001, moved to Namdroling.

I put the question about digital distraction to Damchoe Wangmo. It's wrong to assume that "distraction comes from outside influences rather than inner mental conditions," she explains. If you start with a distracted mind, the ping of your cell phone and the buzz of the Web will tug at that distraction, but they don't cause it. Distraction doesn't come from the outside world to bother a mind that's untroubled. The normal, everyday mind generates plenty of its own distractions.

I put the same question to other monks and nuns, and the responses were puzzling, then fascinating. At first, some didn't understand the question, and I had to explain what I thought was evident, that technology was a special source of distraction. Once I made myself clear, many replied with a question: Why do you think distraction comes from the technology? The whole purpose of mental training is to become immune to such things. "Distractions exist with or without PCs," one monk pointed out. Indeed, Samahita said, external distractions like technologies are "much easier to handle than distractions coming from inside the mind itself."

This attitude makes monks uniformly unimpressed with Zenware (a term they find amusing). "I think the intention of such software is good," Damchoe Wangmo says, "as it presumes an interest in self-control on the part of the user. But it also reflects an incorrect view that distraction is caused from outside influences rather than inner mental conditions."

"Programs and blocks are okay, but ultimately we must build up our own will power," an American nun tells me. "Only we can be responsible for ourselves and what we are doing." "You have to develop your own peace of mind, it cannot be given to you," an elderly monk agrees. "There is no quick fix. You have to develop a practice and apply it daily for benefit to occur." Software can be disabled when temptation is strong, and it can become a crutch instead of a benefit. Sister Gryphon, a Buddhist nun who lives in the Maine woods, explains, "When we see clearly and understand ourselves and our reality, there no longer is a struggle." "Eventually everyone has to take the bull by the horns and subdue the internal noise of the babbling monkey-mind," Bhikkhu Samahita says.

But doesn't all this separation from the world get, well, *boring?* For most of us, the computer in a monk's cell would be the only thing we'd find interesting. What kind of state of mind turns that equation around? What I need to understand is that "renunciation is a trade up," Jonathan Coppola, an American friend of Samahita's, says, "not a trade down." For monks, giving up worldly goods isn't just an exercise in self-discipline or an abstract idea of purity. They give up these things in order to liberate themselves. Freeing yourself from things that don't matter lets you focus on the things that do.

When I understand this idea, the monks' responses make a lot of sense. It's not a matter of resisting the siren song of the Web. Their answers reflect a profound sense that distraction is less compelling than concentration. "How much contentment do distractions yield?" Coppola asks. "If I sit here with my breath and the present moment, isn't that more serene and rapturous than watching cats barking on YouTube? The answer is, of course it is." This is what allows monks to use computers mindfully, to construct practices that express compassion and detachment, and to render the question of Internet addiction or digital distraction irrelevant. A mind like a mirror doesn't need cat videos.

If you see everyday life as full of illusions and the pains of life as the product of illusory beliefs that lead one to misery, if you spend years disciplining your mind and cleansing yourself of false attachments and beliefs and devote yourself to learning how to observe every moment and sensation without assumptions or preconditions, the shiny-blinky Internet, which supposedly appeals to the most primitive, impossible-to-turn-off lizard brain in each of us, is nothing. Few of us can hope to achieve the kind of rigor that these monks do. But we can still apply their insights to our own technology-saturated lives. Buddhists have spent thousands of years studying how the mind works. If they say that distraction arises from within the mind, not outside it, it's worth taking the idea seriously.

It may seem odd that ideas about the monkey mind still seem relevant today. After all, Buddhism developed in India twenty-five hundred years ago. What does life back then have to do with our modern switch-tasking, hyperconnected world? The answer, it turns out, is a lot.

Even twenty-five hundred years ago, shamans, hermits, oracles, and other holy men and women had existed for ages, and no doubt some had developed practices we'd recognize today as contemplative. But by and large, those practices were unwritten or secret, available to only insiders and adepts. By contrast, Buddhist and Taoist instructions on meditation, which date from the sixth century BCE, were widely publicized, strikingly empirical, and meant to turn meditation into a spiritual discipline accessible to everyone. Permanent institutions supporting lives organized around contemplation and spiritual purification soon appeared: Hindu ashrams and Jainist monasteries, followed in the second century BCE by the Jewish Essenes.

Why did contemplative practices flourish in this era? In 1949, the German philosopher Karl Jaspers coined the term *Axial Age* to describe the remarkable period of spiritual and philosophical creativity between 800 and 200 BCE. In this period, he argues, "the spiritual foundations of humanity were laid simultaneously and independently in China, India, Persia, Judea, and Greece." In all these regions, scholars asked profound questions about what it meant to be human; how humans perceive and know the world; and how humans relate to one another and society. As Karen Armstrong put it, Greek philosophers, Buddhist monks, Jewish priests, and Confucian scholars "pushed forward the frontiers of human consciousness and discovered and transcended dimension in the core of their being." The rise of contemplative practices is part of a larger formulation of modern ideas about what it means to be human.

Contemplative practices were a response to the turbulence caused by imperial expansion, political upheaval, global-trading networks, mass migrations, and urbanization. At its worst, life in Warring States China, ancient Greece, or a Middle East contested by Greek, Roman, and Persian empire builders was violent, turbulent, and brutal. In good times, city life offered pleasant diversions and an ever-growing range of distractions. Axial Age philosophers and spiritual leaders advocated rationality and nonviolence as a response to these trials, but they did something more: they reoriented religion itself, turning it away from what John Hick called "cosmic maintenance"—rituals and sacrifices designed to assure good harvests, the smooth passing of the seasons, and so on—and toward personal improvement and enlightenment. One moved beyond a life that was nasty, brutish, and short and ultimately created a better world by cultivating what S. N. Eisenstadt called a "transcendental consciousness": an ability to step away from the world (sometimes almost literally, as with monastic and hermetic traditions, but more often psychologically) and observe it without prejudice or precondition. In other words, to observe it mindfully. To contemplate it.

The Axial Age came to an end around 200 BCE, but the evolution of, on one hand, complex societies, global economies, and empires and, on the other hand, institutions and spaces devoted to cultivating memory and contemplation continued. In the West, the monastery, the cathedral, and the university all evolved into vast machines for supporting and amplifying concentration. These institutions offered an escape from the everyday world, but they also depended on it. Medieval universities in Paris, Bologna, Oxford, and Cambridge raised barriers against the distractions of normal life while at the same time welcoming students, gifts, royal patronage, and high-tech supplies like paper, scientific instruments, and books.

Historians argue that the Internet is merely the latest in a series of irresistible technologies, going back to the invention of writing, that change human brains and the way people think. They've told only half of the story. There's a parallel history of people responding to complexity and turbulence and technological change by creating contemplative practices that support concentration, encourage calm, and restore attention. Worldly distraction and contemplative practices are related: each shapes the other. It's no surprise that ancient contemplative practices should find a resonance in today's superdistracted world. They are tools made for worlds like this and for minds like ours.

FOUR

DEPROGRAM

Think about the first computer you ever owned. Think back to the year you bought it. How fast was its processor? How much RAM did it have? How big was its hard drive? Now do the same thing for the most recent computer you bought—maybe a tablet or smartphone. The difference between the two is a measure of how much personal computers have changed in your lifetime.

If you want a slightly more technical measure, you can easily calculate how many turns of Moore's law stand between the two machines. Moore's law states that the density of circuits on a computer microprocessor—the heart of your PC or device—doubles roughly every two years. So if there are ten years between your first computer and your most recent one, they're separated by five turns, which translates to a thirty-two-fold increase in circuit density. Two years from now, that'll double to sixty-four-fold. Two years after that, it'll double again.

The idea that computers constantly become cheaper and more powerful isn't an abstract notion or science fiction. It's something we've all seen in our own personal history of computing. My kids have already seen big changes in computers. My oldest child has lived through the growth of wi-fi, the appearance of cheap smartphones, Facebook, and five turns of Moore's law. And according to some futurists, by the time she graduates from college, she'll be able to buy a computer whose intelligence and memory will rival her own.

Now think about your brain. Has it followed its own Moore's law–like exponential curve? (It actually did for a little while, but that was before you were born; it also grew dramatically when you were a baby and toddler.) Do you know more than you did when you bought your first computer? Are you able to remember more now? Over the next few years, as computers get even faster and cheaper, do you expect your brain to become faster and be able to store more information?

The answers are probably no, no, no, and definitely not. Computers do things much faster than we do. They continue to improve exponentially. They become more complex *and* simpler, more powerful *and* smaller. We just get older. It's easy to be impressed, and a little intimidated, by the pace of technological change. Living with computers changes how we think about ourselves, our intelligence, and our memories. And those changes have, by and large, been for the worse.

Twenty years ago, Stanford professors Byron Reeves and Clifford Nass made a startling discovery: People treat computers like people. Even the most uninformed user knows that a computer doesn't have feelings or a personality, yet in an elegant series of experiments, Reeves and Nass showed that users unconsciously apply social rules and norms to computers. We assume male-sounding computer voices are more competent than female, particularly when they're talking about technical subjects. We trust computerized "agents" (think of Max Headroom) more if they're the same ethnicity as us. We're even polite to computers; in one study, users who tested a computer program and then completed evaluation forms on the same computer said nicer things about the program than those who filled out their forms on a different computer or on paper. Computers' interactivity and responsiveness make it easy for us to connect with them, much as we anthropomorphize dogs or sympathize with large-eyed kittens.

The fact that we subconsciously think of computers as people makes it easier to compare their rapid progress to our glacial evolution and find ourselves wanting. And as computers become even more responsive and interactive, more fluid and (in some ways) social, their capacity to affect us grows, as does the apparent gap between ourselves and our digital creations.

So if you want to create a better, more mindful relationship with information technologies, you need to understand how computers program us. If you want to get a handle on how technologies affect our images of ourselves, you need to go virtual. Specifically, you need to get into the VR room in Stanford University's Virtual Human Interaction Laboratory.

The lab's director, communications professor Jeremy Bailenson, may be the only Stanford faculty member who's the subject of an iPhone app. An even better measure of his professional reputation is his lab's address. Real estate in Silicon Valley is outrageously expensive, and competition for lab space on Stanford University's vast yet supercrowded campus long ago turned into a blood sport. The VHIL occupies several rooms on the top floor of Stanford's vast Romanesque main quadrangle. The offices of the university president are in an adjacent building. Location matters, even if you spend most of your time in the virtual world.

I meet my guide, Cody Karutz, near the elevator across from the lab. He leads me through an orange waiting room decorated with a 3-D television and copies of *Infinite Reality*, a book coauthored by Bailenson and Jim Blascovich, into the main virtual-reality room.

The lab is one of the world's leading centers for virtual social science, an emerging field that studies how real people interact in virtual worlds and uses VR to better understand everyday human behavior. At first, the VR room just looks like a windowless hotel conference room, all neutral carpet and inoffensive colors. Cody, who combines the all-absorbing passion of a researcher with the easy enthusiasm of a camp counselor (which he was during summers in college), points out the three Kinect controllers and eight video cameras arrayed around the room. Behind the walls are two dozen speakers connected to a sound system that researchers can use to manipulate the room's acoustics and make it seem much smaller or larger. The floor is outfitted with low-frequency transducers that can cause the room to rumble and shake. (I assumed the ramp into the room was to accommodate wheelchairs, but it turns out there's hardware under us.) The closet is a server room packed with eight high-powered graphics-rendering machines and a tangle of cable. I'm reminded of a scene in a spy movie where a character walks through the market stalls in a dusty bazaar, opens a battered door, and steps into a gleaming command center.

In the center of the room stands a tripod topped with a Styrofoam head and a virtual-reality headset. The headset consists of two small high-definition displays, an accelerometer, and an infrared unit that the Kinect controllers use to calculate the user's position in the room. It has the look of a technology that's constantly being tinkered with, yet it costs almost as much as a really nice car. I'm immediately intrigued, and I'm also afraid of breaking it. Wireless headsets can't "update the world"—I love that turn of phrase—fast enough to prevent cybersickness, the sense of vertigo that's induced by the system reacting too slowly to the user's actions, so a black ponytail of fiber-optic cable physically connects the headset to a rack of servers.

The headset switches on. To make things easier for the researchers, a projector throws an image of what the headset shows onto the far wall. They took the architect's AutoCAD files for the room and converted them into virtual-reality code, and now they can add mirrors, new doors, or whatever else they want, depending on the experiment. Right now, the projector shows a picture of the far wall, slightly askew. It then twists sickeningly as an assistant tilts the headset and lifts it up for me to put on.

I close my eyes as she lowers it onto my head and then adjusts the strap that will keep the forty-thousand-dollar piece of hardware from flying off me. When I open my eyes, I'm looking at a near-perfect replica of the room, but Cody has vanished. I hold my arms out in front of me. Nothing. The room is here, but I'm invisible, and Cody is a disembodied voice. What makes these absences notable is that the room looks *really* realistic. Not so much because it's lifelike—it's an accurate rendering, with that artificial sharpness of television-show computer-generated imagery (CGI)—but because the view changes perfectly with my movements. I turn my head, and the door is there, the window to the control room is there, the corners are just where they are in real life. Most people think it's all about high-quality graphics, but what separates the great from the good virtual reality, Cody explains, is tracking. A lot of the reality of virtual reality depends on good tracking. A cardboard-cutout world that you can move in fluidly will seem more real than a beautifully detailed world that stutters around you.

The floor rumbles. I look down, and the area right in front of me is opening up, revealing a deep metal pit with Do Not Jump painted on the floor far below. Of course, I know intellectually that nothing of the sort is happening, but my body thinks I'm really standing on the deck of a carpeted aircraft carrier and looking down into the hangar bay; I feel my heart rate jump and my adrenaline surge. Cody explains that they use this as a demonstration of how real virtual reality can feel, and it works. A small wooden plank extends across the pit, and he invites me to walk across it. I start walking, carefully. I'm halfway across before I notice that I'm holding out my arms to keep my balance.

Bailenson's group has built this lab because people can't help but respond to virtual worlds as if they are real. A traditional scientific laboratory is a miniature, self-contained universe in which a researcher can change one variable in a physical system and observe the results, and Bailenson realized that in a virtual lab, he could experiment with *social* systems. He and his students use virtual reality to create human-looking avatars, but they can change the avatars' voices, genders, races, heights, and just about anything else and then observe how those changes affect human behavior and decision-making. (Other researchers are treating virtual worlds such as Second Life and World of Warcraft as sites for anthropological fieldwork; they're studying how social and economic behavior works when someone is able to become, say, a Yeti or a thousand-year-old magician.) Bailenson and his students have discovered ways to make politicians seem more trustworthy, virtual teachers seem more authoritative and engaging, exercise seem more appealing, and conservation seem more instinctive. They've even figured out how to use avatars to change people's perceptions of themselves.

Some of their first experiments used morphing software to measure the impact of visual similarity in social evaluation. This software allows them to create a realistic-looking image that blends together pictures of different people. In real life, most of us find visual similarity reassuring; people who look similar to oneself are taken more seriously, assumed to be more trustworthy, and considered more attractive than others. Bailenson and his colleagues wanted to know if people would also respond to virtual characters who were made to look like them—and whether they would notice the manipulation. In one experiment, subjects were presented with unaltered pictures of a politician along with pictures of the politician that had been digitally manipulated to include elements of each observer's own face. In a second experiment, members of a four-person group were shown a picture of someone whose face was made of a blend of the group and presented with an argument that they were told came from that person.

What did they find? Participants considered the digitally altered politician more appealing (though partisans were less moved by the changes than the independents). Likewise, groups rated arguments presented by the blended "person" as more persuasive than the same arguments given by a real person who didn't look like a group member. Incredibly, nearly everyone failed to notice these manipulations. Even Stanford's tech-savvy students almost never realized that the image they were shown had been altered. They said only that the politician looked familiar or that he kind of resembled a relative.

Next, Bailenson and graduate student Nick Yee asked if such alterations would work with live interactions. It's more technically challenging to make a virtual animated figure than it is to morph a picture, but these days, with cheap Webcams, facial-recognition software, and speedy computers, it's not *too* hard. Bailenson and Yee developed a system that created a virtual representation of an experimental subject that could be updated in real time, using cameras and image-recognition software to track the subject's facial expressions, eye movements, voice, and so on. The subject sees an uncanny animated version of herself, an avatar that moves when she moves, looks where she looks, and speaks with her voice. It's like looking in a mirror.

In another experiment, they set up a virtual lecture hall, complete with a virtual lecturer. Good public speakers make eye contact with audience members, although a speaker can't look at more than one person at a time. Bailenson manipulated the environment so that, regardless of where the lecturer actually looked, each viewer saw the avatar maintaining eye contact with him, not looking around at different people, down at his notes, or over at his laptop. (Imagine a televised speech by a master politician like Ronald Reagan, a man who had a knack for giving millions of viewers the impression that he was looking each of them straight in the eye and talking to that person alone.) Subjects rated a virtual professor who looked at them all the time as more persuasive than one who didn't—and again, they couldn't detect the manipulation.

Next, the researchers tweaked the system in a couple of creative ways. By switching a few virtual wires, they made it possible for the avatar to copy anyone's behavior. Imagine a mirror that shows not you but someone doing exactly what you're doing. They also altered the system so the avatar didn't mimic everything; it might copy your facial expressions and gestures, but its mouth wouldn't move if you spoke. Finally, they introduced a time delay so the avatar wouldn't copy you instantly but would wait a couple of seconds. They then sat each subject across from a giant screen with a life-size version of an avatar and turned on the cameras. The avatar read everyone a four-minute statement; half the time, it mirrored the viewer's facial expressions and movements after a four-second delay, and the other half, it didn't. The statements read by mirroring avatars were rated as more convincing than the ones read by the less interactive avatars. The results were a brilliant confirmation of the chameleon effect, the theory that mirroring subtly but powerfully affects one's opinion of other people's attentiveness, attractiveness, and persuasiveness.

Decades ago, psychologist Daryl Bem observed that people's attitudes about themselves were defined in part by what they believed *others* thought of them. People who dress in black clothing feel edgy and imaginative because they see people dressed in black in art galleries, and people who wear suits tend to act professionally because they meet people wearing suits in offices. Bailenson and Yee, now a Xerox PARC research scientist, found that when the subject's avatar was taller or better-looking than the subject actually was, he behaved with greater confidence, negotiated more aggressively, and acted friendlier than he did if he was given a short troll avatar. Height and good looks give people confidence because they assume others find them attractive. And it didn't require hours of experimenting with the avatar; Yee and Bailenson saw "dramatic and almost instantaneous" changes in behavior when people saw themselves as taller or prettier.

These behavioral modifications aren't just for that moment—they apply forward too. Other lab members have determined that virtual versions of an individual can change how that person thinks about her *future* self as well as change how she behaves in the present.

When you don't set up the automatic coffeepot before you go to bed, you're doing something for your present self, the person you are right that second. Maybe you're tired and you just want to go to sleep. The problem is that this comes at the expense of your future self: the person you'll be tomorrow will have to make his own coffee. We do this kind of thing all the time—we spend money that we could put away for retirement, watch TV when we should be studying, and so on. Often we know that it's in our long-term best interest to sacrifice for the future, but we find reasons not to do it.

Sometimes the benefits of doing something right now are very clear, while the benefits of sacrificing on behalf of your future self are uncertain. If I spend money now, I get stuff; if I save it, I might have money for retirement—or the economy could tank, and then I'd lose everything. The tangibility of present rewards versus the uncertain value of sacrifices combined with our terrific capacity for rationalization makes us geniuses at justifying living in the moment at the expense of the future. At a restaurant, I might think to myself, *If I eat very little and exercise constantly for the next year, I could look like Tom Cruise. Then again*, I might think as the dessert cart rolls up, *I can go to the gym tomorrow and work off the dessert. Besides, Next-Week Self might give in and eat a whole tub of chocolate ice cream, which would mean that I suffered now for nothing, and Next-Year Self will stay fat. I hate Next-Week Self. No way am I letting him screw us all. What flavor is that cheesecake?*

I might have gotten a tiny bit closer to my ideal weight if I restrained myself, but it's hard for imagined health improvements to compete with real cheesecake. Exercise presents a similar challenge: it's easier to imagine the pleasure that comes from eating popcorn and watching a movie than to imagine the satisfaction you'll eventually derive from being able to run a 10K in forty minutes.

Jesse Fox is studying how virtual-reality avatars can help make those long-term, harder-to-imagine benefits come alive. Her work is full of surprises, starting with the fact that it's conducted by a self-described social-media abstainer who finds books more compelling than video games. "I didn't even own a computer before I started my PhD" at Stanford, she tells me from her lab at the Ohio State University, where she's now a professor. "I've never really liked television either. I'm actually teaching a video game class, and I have to force myself to play LA Noir." She's never had a Facebook account either, and in graduate school, she says without a trace of sarcasm, "I had the *wonderful* experience of being socially ostracized for it."

But she knows how to turn her lack of fanboy fascination with technology into a virtue. It lets her focus on people and their interactions with technology rather than on hardware. And one of her latest research projects studies how being on—or abstaining from—Facebook affects people's social lives.

While she wasn't one of nature's technophiles when she joined the Virtual Human Interaction Lab, Fox was interested in "using fancy equipment to make the world a better place." As a long-time athlete and trainer, she was familiar with the challenges people face when starting an exercise regimen. So for her Stanford dissertation, she engineered a set of experiments to see if virtual feedback could get people to work out more.

She first constructed a virtual room like the one I visited, and she designed a test to see if watching an avatar exercising would get people to exercise longer or more regularly. She created two sets of avatars: one a set of generic-looking figures, and a second set that looked like the students in the experiment. Students in one group put on the headsets and saw their avatar selves running on a treadmill; students in a second group saw their avatar selves standing around; and students in a third group saw generic people running. When Fox followed up with participants, she found that "those who watched themselves running were motivated to exercise, on average, a full *hour* more than the others. They ran, played soccer, or worked out at the gym." Even on a campus where every student has access to Olympic-quality facilities and can exercise year-round, seeing a virtual self exercising had a big impact on behavior.

She next wondered, Could you boost the effect by showing the *results* of exercise or inactivity? One of the problems with exercise and dieting is that it's all pain and inconvenience at first, which makes it easy to give up before the rewards become tangible. Fox devised a second experiment in which participants watched avatars of themselves getting thinner as they ran, or fatter as they stood around—in other words, offering positive and negative reinforcement, respectively, by making the results of a long-term exercise program (or a lack of one) visible in just a few minutes. (A third group watched avatars of others.) "It didn't matter if we threatened you with weight gain or rewarded you with weight loss," she says. "People exercised more when they saw themselves rather than a random avatar."

Why is seeing your virtual self effective? In a final test, Fox put people back in the virtual exercise room, but this time she strapped some exercise physiology equipment on them to measure their heart rates and other vital signs while they watched their virtual selves. "Watching ourselves work has some physiological arousal that motivates us to exercise," she notes. People "actually sweat more when they see *themselves* running" than when they watch others running.

It all sounded a bit like Miguel Nicolelis's cyborg monkeys in reverse. Rather than a subject influencing a virtual self—like the monkey walking on a treadmill and controlling a robot—in Fox's experiment, a virtual self influenced a subject, helping to make uncertain and abstract future benefits more tangible.

Most of us have a very difficult time imagining our older selves. When I was young, I had no trouble imagining myself thanking my supermodel wife during my Nobel Prize acceptance speech, but it seemed ludicrous to imagine myself at age forty. One of Mother Nature's more inspired acts of perversity was to give us minds that are great at conjuring up outlandish things, like my replacing Keith Richards on the Rolling Stones' next tour or piloting an X-wing fighter over the surface of the Death Star, but terrible at envisioning the inevitable, like growing older. As Oxford University philosopher Derek Parfit notes, if we regard our future selves as strangers, we're reluctant to sacrifice on their behalf and more likely to put our present desires ahead of our future needs. In the abstract, I know that when I retire, I'll want to have money saved so I can live well, but it's hard to act on that if my future self is even more abstract.

Fox's studies make the virtues of exercise more tangible, but the virtual selves people encountered in those cases were not older versions of themselves, just more athletic versions. Fox is joining forces with Grace Ahn, a fellow Stanford alum who's now a professor at the University of Georgia, to see how meeting an aged virtual self—the person you'll be in twenty years—can help you make better choices in the present. Their new experiment puts students in a virtual environment equipped with a mirror that "reflects" an avatar whose aged appearance demonstrates the long-term effects of overexposure to the sun. "If you see yourself under the sun, and you see your avatar growing excessively old, we think it will be a powerful intervention," Ahn explains. "For kids who can't even imagine being twenty-one, seeing themselves with a hideous complexion and spots can have an effect on their behavior."

Psychologist Hal Hershfield wants to know if meeting our older virtual selves can encourage us to make choices on behalf of our future selves at the expense of the present ones. Like Fox, Hershfield doesn't really identify as a techie. "I grew up with Nintendo and Super Nintendo," he says, "but I didn't have any immersive gaming experience or immersive world experience" before coming to VHIL. He wasn't exactly a stranger to using interesting technologies to pry open the mysteries of intertemporal decision-making. In his earlier research, he used neuroimaging to compare the parts of the brain used to imagine one's future self with the parts used to imagine strangers. Hershfield reasoned that if Derek Parfit was right and we really do regard our future selves as strangers, then "when thinking of a future self, we should see the same neural patterns as when we think of a future person." It turns out we do.

The results got Hershfield thinking: What would happen if people had *more* vivid representations of their future selves? It's one thing to imagine your elderly self; it's something else entirely to *see* him staring back at you. He mentioned this idea in a lab meeting, "and someone said, 'Oh, there's this VR lab on campus.' So I e-mailed Jeremy [Bailenson], explained my work, and it just took off from there. I was in the right place at the right time."

Hershfield's first study had participants interact with either aged avatar versions of themselves or current versions of themselves. A subject put on the VR headset and saw a virtual room. Once she was accustomed to "being in" the room—some people needed a moment to experiment with the goggles, moving their heads this way and that, to make the illusion seem real—she would turn to a virtual mirror mounted on one wall and see either an aged or a non-aged version of herself looking back. (Hershfield had taken pictures of participants and used an age-progression algorithm to create as realistic an older version of each subject as possible.) The avatar copied the subject's movements and gestures, just as a mirror image would. Next, the subject would watch her reflection as she answered some questions that would "get [the subjects] to identify more with what they were seeing." So, unlike Fox's experiment, subjects did more than just *see* versions of themselves; they were encouraged, however briefly, to *be* their older selves.

Once they came out of the virtual room, subjects were asked a question: If they received a thousand dollars and could spend some and save some, how would they divide it up? These kinds of monetary-allocation tasks are a common way of measuring how much people value their present selves over their future selves: the more money they choose to save, the more future-oriented they are. In this case, it was a nice way of measuring how much becoming one's aged self briefly affected feelings about the future self. What Hershfield found was that "those who'd been exposed to their future selves put twice as much in a retirement account" as those who'd seen their present selves.

But were subjects thinking about their future selves when they made their choices? Or were they just primed to think about old age generally? To answer this question, Hershfield put each participant back in the virtual room, but this time the subject saw either her digitally aged self or another older person. If people had been responding to cues about age, then when the subjects left the room and did the monetary-allocation task again, they should have responded the same way they had before. Instead, people who'd seen their *own* future selves saved more than people who'd seen others.

In a final experiment conducted with psychology professors Dan Goldstein and Bill Sharpe, Hershfield moved from the all-immersive world of virtual reality to the more attention-fragmented World Wide Web. They designed several Web pages that were similar to what employees might use when setting up their retirement benefits; each had a slider bar that participants used to allocate portions of their paychecks for retirement. Two of the pages featured images of either their current or their future selves that would "change emotional expression as a function of their allocation": they'd smile when the allocation was high and frown if it was low. A third page showed either a current or a future self whose expression didn't change with the money allocation, "to make sure people aren't just chasing the smile." A fourth page had the sliders but no face. Once again, people who saw their older selves allocated the most money to retirement; it didn't really matter if the face responded or not. There was also little difference in money allocation between people who saw their present selves and people who didn't see any pictures; the present-self face didn't have a huge effect, Hershfield speculates, because normally "we're already thinking about the present."

I reach the end of the plank in the virtual lab and turn around to cross back again. I have heard of acrophobics who absolutely can't walk the plank and of people who scream and flail as they fall, and damned if I'm going to be one of them. After a couple steps, though, I stumble, and I feel a surge of adrenaline as I struggle to catch myself before I fall into the pit. The rational part of me knows that it's absurd, that I'm standing on a carpeted floor that hasn't changed at all. But my spike in heart rate and my shaking hands say otherwise. I can't help but believe my lying eyes.

Then something else jolts me. I realize it's not quite right to say that virtual reality works because it bypasses your rational brain or because it jacks a digital feed into your nervous system. It's more like theater: it works on me because I let it work.

Grace Ahn explains that virtual reality succeeds because of immersion and presence. *Immersion* is created by stereo sound, naturalistic lighting, detail, and tracking. The physical lab, with its huge array of technologies, is a vast monument to immersion. *Presence,* by contrast, is "a *belief* that this environment is real," Ahn says. "The virtual is nothing but a visual illusion that people are willing to believe." As Jesse Fox puts it, "Presence is all in the mind." It hasn't occurred to me before, but Fox and Ahn are arguing that virtual reality isn't created by technology and consumed by the user. It's co-created by technology and user working together.

"It's not that people don't understand that it's not real, but it's realistic enough to willingly suspend disbelief and embrace the idea that it's real," Ahn says. At some level, "you understand it's all simulation, but you're willing to accept your virtual experience as real enough to allow you to learn something from it."

A user's prior experience, his ability to focus on the digital moment, and his habits of concentration all shape how much presence he experiences, or doesn't. After some experiments, Fox says, "I've had people come in and say, 'VR sucks, man. Call of Duty is *so* much more awesome.' " For gamers, the feeling of really being in a virtual world hinges on their being absorbed by a task. I immediately understand what her gamer critics mean: I play Mario Kart on my kids' Nintendo Wii, and I'm very good at it because I can completely block out the busy surroundings and focus entirely on the road in front of me. Virtual reality is powerful when eyes, mind, body, and technology are profoundly entangled. The technology provides the virtual, but we provide the reality.

Virtual reality offers an example of how powerful entanglement is when it goes well. When it goes badly, entanglement can lead you to question your own intelligence.

One morning my iPad and Bluetooth keyboard stopped communicating. Connecting them should have been a simple matter of making sure that the keyboard was on (there's a power button at one end of the keyboard), then switching on the iPad's Bluetooth and telling it to connect to the keyboard. Normally, they find each other, but this morning they weren't connecting at all.

I couldn't be sure where the failure point was. If it was in the keyboard, was it the batteries? The antenna? A mechanical problem? The keyboard couldn't tell me anything; it has a single tiny green light (actually a very small laser) that glows (fires) when you push the power button. You can see if it's on or off, but it can't tell you if it's broken. I replace the batteries and restart the keyboard. Still no luck. I take the new batteries back out and test *them,* just in case. They're fine. So maybe it's a problem with the antenna? But wait—which antenna, the keyboard's or the iPad's? I get out my iPhone (yes, I know) and switch on the Bluetooth connection, and it finds the keyboard. That suggests the problem is with the iPad. A few seconds later, the keyboard disconnects from my iPhone. Inconclusive.

I spend the next couple minutes pressing the keyboard's power button and watching it pop up in my iPad's list of available devices and then disappear again. I briefly wonder if there is something in the environment that is interfering with the signal. But I have no idea how to detect such an interference, much less deal with it. I might as well have been considering the possibility of gremlins.

I don't have time for this, I keep thinking, a thought closely followed by variations on *Technology is stupid* and *I must be missing something really simple, because this has always worked before.* At the same time, I can see that the episode illustrates how technology failures make *the user* feel stupid. Until you solve the problem—or, somehow worse, until it spontaneously, mysteriously fixes itself—you don't know enough to be sure where the source of the error lies: in a device, in a subsystem, or in yourself.

Problems like these turn into frustrations because information technologies are opaque. Companies want their products to look easy to use, but too often, that means immense complexity hidden behind brushed aluminum and smoked glass. That makes it harder for you to understand what's going wrong, deduce what kind of problem you're dealing with, and learn enough so you can fix it the next time.

Opacity becomes more problematic as technologies become more complex. Yale sociologist Charles Perrow has argued that catastrophes such as the *Challenger* shuttle disaster aren't caused by shocking lapses of professional judgment or once-in-a-lifetime glitches; in his terrifying term, they're "normal accidents." Tightly coupled, highly interdependent systems fail catastrophically when small changes and everyday errors cascade unpredictably into massive problems. My keyboard's sudden inability to talk to my iPad wasn't a big issue, but when you live surrounded by information technologies and spend your time interacting with them or interacting with the world through them, you're perpetually exposed to these minor, annoying emergent misbehaviors.

Interruptions like these aren't necessarily harder to deal with than technical problems you deal with at other times, but because they break the flow of your work and concentration, they're painfully memorable. As computer scientist Helena Mentis discovered, people remember problems that occur when they're trying to finish up a project—a slow response to a command, a pop-up that distracts them from a Web page, or other examples of what interface designer Alan Cooper calls "stopping the proceedings with idiocy"—better than they remember problems that occur at other times. Your response to a problem that occurs at the beginning of a project will be very different than your response to a problem that occurs when you're much farther along in the process—say, when you've been working for hours against a looming deadline, finally finish, hit Print... and nothing happens.

(An hour later my keyboard suddenly started working again. I don't know why.)

Computers have other ways of making us feel dumb. We've long associated intelligence with speed. In English, we say that someone is "a quick study" or a "fast learner"; it's not a compliment to call someone "slow." By this standard, computers have a decisive advantage over humans. They can effortlessly and almost instantly do things that we have a lot of trouble with. They're already fast, and they're getting faster, cheaper, and more powerful, while we're still stuck with virtually the same type of brain that our ax-wielding, cave-dwelling ancestors had. The future of computer intelligence seems boundless, while the future of human intelligence (without the aid of genetic engineering or brain-boosting drugs, anyway) looks limited.

We're told that new forms of digital intelligence are evolving. Swarming robots and computer programs reveal the power of emergent behavior, the ability of large numbers of semi-intelligent agents to create very intelligent systems. Crowdsourcing, Wikipedia, and prediction markets demonstrate that an entirely new kind of collective intelligence, global in its reach and literally superhuman in its potential, is emerging on the Web.

Virtual-reality pioneer Jaron Lanier argues that in the course of pursuing such technologies and crowdsourcing projects, we've accidentally recalibrated our ideas about human work and worth. "People degrade themselves in order to make machines seem smart," he says. For example, Amazon's Mechanical Turk, which lets companies farm out to freelancers large numbers of small, simple tasks—such as describing the contents of an image, for example—seems to imply that what's valuable about humans are their numbers and flexibility, not their intelligence. Wikipedia obscures the identities of authors, Lanier argues, "in order to give the text superhuman ability" and to create the illusion that knowledge emerges spontaneously from the collective. Systems like InnoCentive, in which companies can offer bounties for solutions to technical problems, are often described as tools to harness the "wisdom of the crowd," rather than as marketplaces in which companies with unusual problems can find experts with unique knowledge.

This recalibration also appears in discussions of the future. Ray Kurzweil's *The Singularity Is Near,* for example, argues that we'll soon have computers that match or better human intelligence and that will form networks of "nonbiological intelligence" whose power will "vastly exceed biological intelligence by the mid-2040s." To those who say that computers can never think like humans, Kurzweil replies that while Shakespeare's *Hamlet* and the Beatles' *Rubber Soul* are pretty great, most "human thought is derivative, petty, and circumscribed." Further, "biological human thinking is limited" by the constraints of the human brain, with its "very slow interneuronal connections." We can tinker around the edges with genetic engineering, but the real future lies in brain-machine interfaces. Our children and grandchildren, he says, will have nanotechnology robots swarming through their brains, connecting their biological central processing units to their silicon and spintronic cousins. They'll offload memories to the Cloud as naturally as we commit them to paper now; a visit to virtual environments will be as real as a trip to the corner store; and merging one person's lifetime memories and feelings with another person's will be just another form of conversation. Our descendants will merge "biological thinking and existence" with the latest technology, but they won't experience the singularity as a massive, wrenching alienation. Humans won't look around and watch it happen. They'll look inside and feel it happen.

Gordon Bell predicts that "electronic personal memories" will transform the way we remember. For nearly ten years, Bell has been the subject of the MyLifeBits project, an experiment that aims to capture virtually every waking moment of Bell's life. He scans or photographs documents he encounters and wears a small camera called a SenseCam that takes pictures throughout his day. The database stretches back to include material from his early life in small-town Missouri during the Depression, the family appliance store where he first learned how to work with electronics, his years at MIT studying the then-novel field of computer science, his two decades at Digital Electronics Corporation, and, most recently, his work at Microsoft Research, which he joined in 1995.

"Most of us are well along the path of outsourcing our brains to some form of e-memory," he and colleague Jim Gemmell wrote, and MyLifeBits aims to figure out what this will mean. Webcams, cell-phone cameras, video cameras, audio recording, and location-sensing GPS are all stunningly cheap, making it possible to document virtually every second of a person's day. Storing all this data is becoming so cheap, it's almost more expensive to throw something away than to keep it. And digital memory is amazingly accurate. Compare your memory of an event to a recording of it, and you'll likely be depressed at how much you've forgotten: *Oh, was she there too? I don't remember the band playing that song*, you'll think. Or, a different but equally depressing comparison, compare your memory of your recent bank or credit-card activity with your online records. You'll find there's a computer somewhere that knows the details of your financial history a lot better than you do. Computers remember passwords that you forget and appointments you don't recall making; they can remember random strings of letters and numbers as easily as you can remember your own name. There's really no contest between stable, digitally sharp computer memories and fragile, forgetful human ones.

But these examples obscure the fact that computer and human intelligence are actually different enough to make direct comparisons tricky. For one thing, there are lots of kinds of intelligence. The ability to recognize faces, detect patterns, use language, navigate social situations, respond to emotions, and do a thousand other things are all forms of intelligence, but they're all a little different, and they use different parts of the brain. The correlation of intelligence and speed is also imperfect. Artists can spend months on paintings (sometimes agonizing over those details that give a work that spark of freshness and improvisation), writers can work for years on books, and scientific theories are built on a mix of lightning-fast inspiration and decades of research. Finally, human intelligence isn't as constrained by biology as Kurzweil would have us believe. We don't get smarter only through physical changes to basic brain structure; we get smarter through cultural evolution too.

Human memory is just as complicated. *Memory* encompasses a wide variety of mental processes: it describes a person's ability to remember words he heard five seconds ago, the place he left his car keys last night, the name of someone he met two weeks ago, the feel of a lover's skin during an afternoon on a tropical beach, the words of his native language, the winner of the final in the 1988 UEFA European Championship (Netherlands over USSR, 2–0), whom to ask to find out which restaurant in Chinatown has the best dim sum, his years in college, and so on and so on, ad infinitum. Short-term memory, visual memory, transactional memory, episodic memory, long-term memory, declarative memory, facts, events, nouns, verbs, traumas, impressions, pictures, feelings—they're all things we remember. They all involve the past, and we use the word *memory* to describe recalling them, but that's where their similarity ends. And while we're impressed by people who are able to remember an encyclopedia's worth of facts, most memory isn't just information in brain storage. *Memory* is a process as much as a thing, a way to reconstruct the past as well as a stable pattern of information saved for the future.

The flexibility of memory is not a bad thing. It allows us to revisit events, to take more mature views of our pasts, to make sense of our lives as we live them. Forgetting can be constructive too. For someone who's recovering from posttraumatic stress disorder, the ability to *not* remember the details of a buddy's death, the ordinary ambiguity and brutality of counterinsurgency, or the terror of being in a damaged plane as it falls over enemy territory is a triumph. In everyday life, the ability to forget the raw feeling of insults and intrusions, the pain of slights and embarrassments, allows grievances to heal. A person who still holds a grudge because of a misunderstanding twenty years ago is not admired for her amazing memory; she's more likely to be regarded as childishly incapable of letting something go.

There's also a social dimension to collective memory and forgetting. In our everyday lives, transactional memory lowers the cognitive cost of remembering because we can outsource the work to other people. Nations choose what events will live on in public memory: fragile states may decide to avoid deep reflection on wars or civil strife in the interests of promoting national unity and moving forward. Having grown up in the capital of the Confederate South, in the shadow of monuments to Civil War generals, I'm aware that what's unforgiven, what's considered not worth remembering, and what we can be strong and generous enough to forget are all subject to negotiation, and what lives in memory very much shapes the present. Some forgetting even has a legal dimension. For a long time, the actions of youths have been understood to be separate from the actions of adults, and even adult crimes can be forgotten, in a sense: someone who does jail time for a crime is believed to have paid his debt to society.

The persistence of digital memory risks short-circuiting these processes. Computers are indiscriminate rememberers. This is an excellent property when it's used to keep track of bank records or subatomic events, but it's problematic when it's applied to the world of complex human affairs. For humans, forgetting can be psychologically valuable, even generous. For computers, it's an error.

Some of the most poignant e-memories that Bell and Gemmell discuss show how digital records are to human memories what a match is to a bonfire: a spark, not the thing itself. MyLifeBits can call up content from virtually any part of Bell's life—everything from his childhood birthday parties to business cards of people he met at a conference. He describes what happened to him once when his screen saver pulled up a picture from his fourth birthday. Even decades later and thousands of miles away, the photo inspired an "avalanche of memories": the cake, the children he invited, the neighbor girl he secretly liked, the minister's son who died tragically soon after. It's positively Proustian. Bell would not have had the photo drop into his everyday world were it not for lifelogging, but what's notable is where the memory he describes so vividly originates. Like Proust biting into his madeleine, Bell looked at the picture and experienced a flood of associations. But the madeleine did not contain the memory of Proust's childhood mornings in Combray, and the photograph did not contain Bell's memories; it called to mind memories that were contained within him. The photograph is not a human memory; it is an image stored in computer memory, and it requires Bell's own brain to give it meaning. The e-memory can help keep those human memories alive, but that's different from replacing those memories.

In the future, Bell and Gemmell speculate, software will be able to create avatars that digest people's lifelogs, build psychological profiles of them from that material, and replicate their personalities. The idea of building avatars based on lifelogs raises as many questions as intriguing possibilities. Which "me" would this avatar be? The person my spouse sees, the person my parents deal with, or the person my boss knows? Would it be the twenty-year-old version of me, the forty-year-old version, or the five-year-old version?

Rather than thinking of digital and human abilities as competing, think of them as complementary. You can remember in ways that computers can't, and they can store and retrieve information more accurately than you ever could. You have the capacity to deal with ambiguity, to make novel connections, and to imagine, which is something no computer can rival; they can work with a precision and attention that you could never match. You can create an extended mind that's strengthened by the joining of diverse skills, not weakened by unending distractions, unhelpful complexity, and unexamined habits.

Computers change our thinking about ourselves indirectly. Occasionally, though, they're designed to shape their users. In some cases, Stanford PhD Morgan Ames explains over vegetarian pho and spring rolls, computer scientists' autobiographies "have a material effect on computers." The two of us are meeting at a trendy restaurant in downtown Palo Alto (the kind of place where angel investors and young CEOs have expensive conversations about search algorithms and disruptive innovations) to talk about how designers' life stories affect the ways they shape technologies. The place is crowded, but Morgan navigates through the tables gracefully. No surprise; her mother is a dance instructor, and Morgan herself was a nationally ranked competitive ballroom dancer before she retired to focus on her dissertation on the One Laptop per Child program.

The brainchild of MIT impresario Nicholas Negroponte, OLPC aims to spark a worldwide revolution in learning. Its flagship product—as practical as an AK-47, as eloquent as the Declaration of Independence—is the XO, a laptop that's cheap and rugged enough to be used by children all over the world. It's robust enough to be dropped by parachute and so easy to repair that a kid can fix it herself. The laptops will go straight to kids; Negroponte argues that by putting the computers directly in the hands of children, kids will be free to explore the machines without the constraints imposed by teachers and formal structures. As a result, they'll learn to program, and they'll unlock the transformative potential of computing.

OLPC reflects a whole set of assumptions about how people learn and how people learn to use computers, Morgan argues. At the program's heart is a belief that great programmers are self-taught, not trained by others. *When I was young, I was bored in school,* many computer scientists and entrepreneurs will recall. *I didn't have many friends. I was an aimless misfit. Then I discovered computers, and it was like a whole world opened up for me. When I sat down and turned on the machine, I became my true self: I learned how to program, I became self-directed, I found something I was passionate about. I didn't let my school's indifference stop me; if anything, I learned that constraints are just new problems to solve. Through those thousands of hours close to the machine, I made myself a programmer.*

What's striking about these stories, Morgan says, is not the message that "computer scientists are made through individual open-ended exploration of computers"; what's striking is what's missing. Many of the hackers she's interviewed had parents who were scientists or engineers, but the hackers often insist they didn't really learn anything at home. And "even if they excelled in schools," Morgan says, they frequently remember "hating it and opposing their teachers." Some even hacked into school computers; Microsoft cofounders Bill Gates and Paul Allen, for example, broke into their high school's mainframe to give themselves unlimited programming time. (When they fought their schools, they were fighting for the right to learn.) "Everything important happens between the engineer and the machine" in these stories, Morgan says. "Family values, home values, modeling what learning should be, how to learn — these all disappear." What emerges is a vision of the programmer as completely self-made and self-reliant.

This solitary-hacker ethos pervades OLPC, and the XO is designed to make kids technically savvy, self-starting, and contemptuous of tradition and authority—in other words, to make them hackers. XO is not a media consumption or communications device. Web browsing, movies, and music were deemphasized in favor of tools like Turtle, in which a user creates art by programming a turtle to crawl around a canvas, and LOGO, a programming language developed by MIT professor Seymour Papert in the 1960s. The early strategy of giving XOs directly to the children rather than schools reflects a deep faith in the ability of all children to learn to program and an equally deep faith that teachers and schools would just get in the way. The belief in self-directed learning and self-made hackers has also led to a policy that minimizes investment in education programs, teacher training, and repair facilities. (The computer, meanwhile, has proved less robust and harder to repair than designers wanted, and even in the Third World, kids use the XO to go online and play games.)

What's striking about Morgan's story is that the OLPC designers don't believe that the power of computers alone is enough to change the world. In Morgan's telling, OLPC is on a mission to spread a certain kind of relationship between computers and children. It emphasizes experimentation, tinkering, a sense that anything is possible with computers, all of which can be good. But it also encourages children to believe that institutions are, at best, irrelevant to learning and, at worst, impediments to it.

Beliefs about ourselves and others can have a powerful effect on human performance and behavior, in part because the beliefs are subconscious. In a classic series of studies, sociologist Claude Steele showed that African American students who were asked to identify their race at the beginning of a standardized test scored lower than classmates who were not; racial self-identification, he argued, triggered a stereotype threat about African American intellectual inferiority. (Similar studies showed women's scores on math tests dropped when they were asked to identify their gender.) Students who think they succeed because of raw talent treat failure as an event that establishes the extent of their abilities. When they're exposed to research showing that intelligence is flexible rather than fixed and that practice is a more powerful determinant of success than innate ability, these students learn to treat failure as a challenge rather than a defeat, and their long-term success improves. Knowing about such effects, though, can let subjects resist them.

It's just as important for us to be aware of how computers have programmed us, of how our models of ourselves have been shaped by our interactions with information technologies. Computers have changed our understanding of *human* work intelligence and memory and led us to value (computerlike) qualities such as efficiency, speed, and productivity more than creativity, deliberation, and thoroughness.

Thinking that intelligence is fixed rather than flexible, believing that our own abilities are but dim versions of those of our digital creations, and accepting that the future belongs to our devices rather than ourselves likewise has real-world consequences. Conflating human and computer abilities and comparing histories and prospects leads to a kind of tweety, blinky despair, a vision of the future that features endless upgrades and perpetual inadequacy, avalanches of novelty, and constant distraction.

Realizing how computers have programmed us can break that cycle of conflate, compare, and despair and help us see how to use computers more thoughtfully. It encourages us to evaluate human and digital abilities separately and on their own terms. Recognizing that human brains and computers have complementary abilities can help us move from comparing ourselves with our devices to experimenting with uses and practices that blend the best of biological and artificial abilities—practices that create extended minds, not distracted ones. The point of such experiments is not to replace our abilities, but to augment them. We needn't accept the idea that a future in which computers are smarter than we are and in which we oversee machines that hold our memories and think with and for us is inevitable. In other words, don't resign. Redesign.

FIVE

EXPERIMENT

Open your e-mail program. Go ahead. You're going to sooner or later anyway.

(Did you remember to breathe?)

For the next few days, go about your normal online business but pay a little more attention to how you interact with your mail. Start with the number of interactions you have with your e-mail program. Write down how many times a day you check your mail, and how many times you get an alert telling you that a new message has arrived. (Don't forget messages on your smartphone.) Take note of where you check your mail—at work, in the car, the kitchen, during the boring parts of the reality-TV show you're secretly addicted to, in the bathroom (be honest). Next, estimate how much time you spend reading, answering, and writing mail. If you have a stopwatch, great, but the results don't have to be precise, and recording them shouldn't be so burdensome that they distort your normal habits.

This time-and-motion observation is the sort of thing that industrial engineers and product designers do. It can be a very illuminating exercise. It often reveals how mindlessly we use computers and the Web in our ordinary lives and frequently demonstrates that we spend more time than we realize with our hands on keyboards and our eyes on screens. Ohio State University professor and virtual-reality researcher Jesse Fox uses a technique like this in her communications class. She has students keep media diaries in which they track every engagement they have with social media, television, video games, and so on. The tracking can get overwhelming—"It's so much work!" they complain—but the results are often shocking: students find that while they *think* they're jumping online just for a minute or having a quick round of Drawsome with friends, they're actually spending two hours a day on Facebook or thirty hours a week playing video games.

What really gets Fox's students thinking, though, is this: Before the experiment, Fox asks them all to make lists of everything they'd like to do but feel they don't have time for. The lists have to be detailed, to get them thinking seriously about how they'd spend their time: an hour every three days doing Pilates, four hours a week to visit someplace new, half an hour every day for coffee with friends, twenty minutes a day for laundry and cleaning, and so on. She then has them compare the diaries with the lists. In the eight hours you played video games, she points out, here's all the *other* stuff you could have done.

But in this case, measuring sessions and usage time is just the start. Because, unless you live on the set of a retro 1960s television show, e-mail is actually important. For years, people who grew up in a world of more formal business communication have been lamenting that e-mail is more informal than memos or real letters, claiming that the time required to write out (or dictate) a letter gave a writer a chance to think about his words and that instant communication promotes sloppy thinking. Maybe. But it's here to stay, and it's important to do it well—for the sake of the people you're communicating with, for your own professional presentation, and for yourself.

You need to move beyond the efficiency and lifehacking questions, beyond thinking about how you can write faster or drain your in-box. Your colleagues don't need *faster* replies; they need *better* ones.

So don't just collect numbers on times and time. How many truly important messages do you get in a day, messages that you need to act on quickly or that give you useful information? How many spam messages do you get? (Check your filter, then empty it. That multimillion-dollar transfer from a Nigerian commercial bank is meant for someone else.) This will give you a better sense of how much time you really need to devote to e-mail.

Finally, consider your attention level and emotional state before and after you check your mail. When you start your mail program, are you anxious because you expect something important? Are you checking it because you're bored? Because you haven't checked it in the last ten minutes? Do you even *have* a reason, or is it just a reflex, what you habitually do when you come to a red light or stand in line for coffee or can't seem to get started on a project?

And how do you feel after you've checked? Better than before? Does your mood improve after checking your mail? How long is it before you start to wonder if there's a critical message in your in-box waiting to be read?

After a week of observing the way you use e-mail and noting how you feel when using it, it's time to make sense of the data and act on it. You spend x number of hours a day on the computer, check your mail y times, and read z messages that matter, and this improves your mood n percent of the time.

Do the numbers you get seem high? Let's figure out how to lower them.

Start first with the emotional data. Are there times of day when it feels most satisfying to check your mail? If there's a pattern, your first step is clear. Try a few days where you check your mail only at those times and no others. Stop "just checking in" when you're at the grocery store or waiting for the elevator; put your phone at the bottom of your bag if you have to. Continue recording information about how many times you check, how much time you spend, how many of the messages are worth your attention, and how it feels to deal with them, then see how those numbers compare to your first set.

If it seems like simply spending less time on e-mail would be good, try setting a couple of specific times a day for you to check your mail. Do it on one machine only; this will cut down on the instinct to use whatever machine is at hand and save you the trouble of having a draft of that critical response on the other machine.

Redesigning your e-mail practices involves a bunch of small observations and experiments, but the result is fairly grand: you are redesigning your extended mind. It requires observing your everyday practices, tinkering with your devices and altering how you use them, and then making conscious choices about how you will use your technologies in the future. In other words, it requires self-experimenting.

Self-experimentation is the systematic observation of one's own physical and psychological reactions to particular stimuli or events. It used to be frowned upon by scientists, considered a narcissistic, subjective, and unreliable research tool. But it has become more legitimate and popular thanks to cheap, easy-to-use monitoring devices that generate precise, unbiased data; flexible analytical tools that quickly find patterns in large data sets; and the growth of online communities that need to find specific, personal solutions to complex problems. Self-experimentation has been used by Nobel Prize–winning scientists (most recently Barry Marshall and Robin Warren, who discovered the bacterial cause of peptic ulcers), people living with chronic illnesses, and elite athletes.

Because it requires paying lots of attention to technologies as well as to yourself and observing how the technologies work—and how they work with you—under a variety of circumstances, self-experimentation provides a fine way to discover some of the subtler benefits and costs of new technologies, reveal the unanticipated ways they affect your mind, and find the unexpected—and sometimes very valuable—skills they can help you develop.

We all have different capacities for attention, different things that distract us, and different things that help us concentrate. It's essential to figure out for yourself what systems work for you. So it's important to tinker thoughtfully with your technologies, usage practices, and work habits in order to learn what combinations support contemplative practice.

The first thing to observe is that it's possible to use familiar technologies more mindfully. Sometimes you can cultivate this facility. Sometimes it sneaks up on you, and the best you can do is be prepared for it.

It hit me one afternoon in Cambridge during a long walk with my wife and a fellow visitor at Microsoft Research. We were trekking from Cambridge to Grantchester to visit the Orchard, a teahouse on the banks of the Cam, the fabled river that winds through Cambridgeshire's fens. For a century, people have followed this path, blazed by the great poet Rupert Brooke when he sought to escape Cambridge, which he described as "urban, squat and packed with guile." While Brooke found "peace and holy quiet" in Grantchester, he actually didn't go very far, as the village is only a couple miles upriver. It's roughly the same distance Henry David Thoreau's little cabin beside Walden Pond was from the town of Concord, my wife reminded me. Contemplation can be closer than we think.

To get to the Orchard, you follow the Cam south through Cambridge to Grantchester Meadow. The gently rolling meadow is quietly beautiful; farm and woodland are visible in one direction, and the spires of the colleges occasionally visible in the distance. It doesn't seem like it's changed much since it was purchased by King's College in 1452, the same year Leonardo da Vinci was born and the first Bible rolled off Johannes Gutenberg's printing press. You might see a herd of red poll cattle grazing near the water. After a mile, when you reach Grantchester village, walk down a narrow walled lane, pass a church, and you'll come to the Orchard.

I thought it was exactly the sort of trip that was meant to be photographed exhaustively, and I'd brought our digital SLR and a couple of lenses. My father was an avid amateur photographer when I was young, but it wasn't until the invention of digital cameras and the arrival of my children that I became a photographer myself. For me, the camera can easily become an intrusive, unwelcome object that distorts the events it's documenting ("Everybody stand together for a picture!") or threatens to pull my attention away from the environment and onto itself. It's easy for me to become caught up in the geekier aspects of the craft. I've spent as many hours experimenting with the virtual lenses and films in Hipstamatic, the iPhone's retro camera app, as I spent mastering mazes in Pac-Man in my youth, and I love the feel of a great camera's heavy precision. Digital memory lets you take vast numbers of pictures, which leads you to believe that by sheer random luck, some of them will turn out well. This attitude encourages profligacy at the expense of skill; speed instead of seeing; distraction and disruption rather than concentration. Why actually frame a picture when you can play the lottery and assume that if you take enough pictures, *one* of them will look good?

Certainly there have been times when I've been too focused on documenting the moment to be *in* the moment. But as we walked through

Grantchester Meadow, I was struck by how carrying a camera encouraged me to look more closely at my surroundings, pay more attention to movement and light and shadow, notice reflections in the water and the contrasts among different shades of green and brown (it wasn't very colorful in the winter).

It wasn't about scanning the environment for good pictures. I realized years ago that the instant feedback a digital camera provides was slowly sharpening my ability to see and frame good pictures, closing the loop between what I saw and how well I captured it. Now, though, I felt the camera receding from my attention, and I found myself instead engaging with the *place.* The pictures seemed to emerge in the course of my watching the shadows against the grass or noting the shades of gray in the clouds. I was reminded of Eugen Herrigel's account of Zen archery, of the years he spent trying to get his shots to fall from the bow like petals from a cherry tree. I was aware of the possibility of finding a little of this casual, effortless attitude with the camera.

The blogging Buddhist monastics show that it is possible to use any technology mindfully. I later discovered another famous monastic who turned photography into a form of contemplation: Trappist monk Thomas Merton, author of *The Seven Storey Mountain* and probably the most famous Christian monastic of the twentieth century, discovered late in life that he could use the camera as a tool for "reminding me of things I have overlooked, and cooperating in the creation of new worlds." It struck me as a great example of how a well-used technology can encourage an attitude (as Merton put it) of "being open and receptive to what comes to the eye," of mindfulness rather than distraction.

As a cow ambled nearby, I wondered, *Where did this come from?* Was this sort of focused attention an artifact of my engagement with the camera? Certainly I was paying more attention because I do so much picture-taking, and I doubted I would have become attuned to details—reflections, shades, the texture of the mud along the cow path—if I didn't. So was the technology training me to look at the world in a particular, more camera-like way? Hard to say. If I'd studied drawing or watercolor painting, I probably would have noticed different things; the Victorian art critic Philip Hamerton argued that watercolors teach you to see the world in terms of spaces and tones, while the pencil encourages you to see it in terms of hard lines and shadows. But any technology—even eyeglasses—has its biases, and all perception has an inherent incompleteness; vision, as scientists have discovered, dances between greedy consumption and smart filtering. There's no seeing without selection.

In the hands of paparazzi and photojournalists, the camera, for better or worse, is a tool for intruding on events and lives. But it can also be used as a reminder to remain, as Thomas Merton put it, "open and receptive to what comes to the eye," a way to engage with the world rather than be distracted from it.

Like any really good example of entanglement, my connection with the camera was encouraging me to push beyond my naked, unassisted abilities, to see with greater care and skill. I could now see more with the camera than without it. The camera had improved my attention, not become the center of my attention. It presented me with an opportunity to be more mindful of my environment, to engage more thoughtfully with my own vision and the world I was seeing. My extended self could be an expanded self.

A very different kind of opportunity to find a more mindful way of using technology, a way that opens up opportunities for flow (that form of deep engagement that improves your facility to concentrate) presented itself a few months later, when we returned home to California: I started playing Mario Kart on our Nintendo Wii.

For those who've never seen it, Mario Kart is a racing game; you use the Wii controller as a steering wheel, and there are simple controls to speed up, brake, or throw things at other drivers. Our family plays a lot of Mario Kart. We try to reach into the shelf stuffed with board games as often as we reach for the Wiimotes, and this has become a regular ritual in our household, but we race one another several evenings a week.

I'm not one of those parents who worry too much about their kids playing video games. I don't want my kids to play them instead of reading or getting exercise, or to play them to the point of being unhealthy or socially isolated, but video games and I grew up together. The first video game I ever saw, Pong, *was* the first video game. I spent thousands of hours of my adolescence playing Defender, Xevious, Missile Command, Star Wars, and Battlezone. When I spent a summer in Japan in the early 1980s, I felt like a jazz fanatic who'd just arrived in New Orleans: I was in the creative center, the home of Nintendo and Namco, the place where Pac-Man was born (or hatched or materialized or whatever). For me, video games were a part of growing up, so it's no surprise that my kids like them too.

But I also recognize that there are downsides to gaming. Too many of the hours I spent in the arcade were occupied in a mind-numbing way, my going through motions I'd learned a thousand games before rather than pushing myself to do something new. Video games can be a narcotic, but so can books. I don't want to forbid either one; I want my kids to be savvy, enthusiastic readers *and* gamers. Playing with them is a way to make that happen; it's also a way to make the experience social, to enforce good sportsmanship, and to not-so-subtly teach the kids how to play well. They have a teacher for a mother, and a father who explains that an ATM returns the bank card before dispensing money to avoid postcompletion errors, so our kids are used to having perfectly ordinary experiences turned into lectures and experiments.

We play Mario Kart because everyone likes it, but also because it's reasonably simple and actually rewards skill and practice. Some games are so complex, you have to take a graduate course to understand all the controls and objectives. You can learn the basics of Mario Kart in a few minutes. Other games reward frantic button-mashing and aggressive, bet-the-house play that causes bad feelings between contestants and headaches in parents; we've banned those. With Mario Kart, it pays to play it cool, stay alert, and react instantly. It teaches the virtue of cultivating persistence and ability.

Mario Kart also offers an opportunity to learn the virtues of total concentration. When I first saw someone playing the game, it seemed incredibly noisy and busy, everything I didn't want in a sensory experience. However, when I tried it, something amazing happened: my mind quickly filtered out the backgrounds, the crowds, the signage and sound effects. Instead of being distracted, my attention was focused intensely on the road and the cars right in front of me. The visual field literally narrows when a person is stressed or concentrating hard, and it impressed me that I could *see* that happening as I played.

This was something I wanted to point out to my kids and that I wanted them to experience. If they could develop a taste for it here, I reasoned, it was more likely that they'd seek it out elsewhere and cultivate it in other things they did. Getting them into that zone, though, was a bit of a challenge. When we first started playing, my son would chatter as we played, and Not. Stop. Talking. Not only is that bad manners, but it messes with my groove. For the thousandth time, I explained that playing with others was like going to the movies with people: you have to respect their desire to focus on the action. Great players, I continued, win because they learn to concentrate hard enough to play well; for really great ones, the concentration becomes its own game, even more rewarding than the one on the screen. My son sometimes still chatters, but he's getting better, and every evening he and his sister get an example of the rewards of focus—and if the parents are on their game, a lesson in how to be graceful losers.

It's not exactly a week in a Zen monastery. But it's a fun way to learn the benefits of maintaining concentration and calm, to see that the way to win any game is to play mindfully, and to find games that support flow. And for me, the challenge of figuring out how to explain this in a way that didn't sound too abstract forced me to pay more attention to why I liked the game, to how I play, and to how playing socially can be very different from playing alone. Explaining mindful gaming made me a more mindful gamer.

Self-experimenting can help you find ways of being more mindful with information technologies. It can also help you be more thoughtful about how your work and cognitive habits are affected or enhanced or degraded by different kinds of technologies. For example, it can help you make the best decision when it comes to one of the most common everyday choices people make: whether to use books, paper, and pen or e-books, screens, and keyboards.

There's been a lot of debate over the future of books and reading, but its focus on cultural institutions like literature and libraries have distracted us from appreciating how tightly entangled readers and texts become, and how often that entanglement is triggered. We live in a world supersaturated with words. We see and react to words on street signs, boxes, newspapers, clothes, dashboards, magazines, and a million other places. (The books category includes catalogs, atlases, histories, encyclopedias, instruction manuals, plays, children's books made of washable materials, and so on.) Our interactions with words are as varied as the places and contexts that we encounter them; we glance at street signs while driving, browse the front pages of newspapers at breakfast, scan Web pages at work, lose ourselves in novels on planes, read our children to sleep at night. Reading encompasses a vast range of activities and is influenced by place, intention, and media.

While many people report they are spending less time reading books for pleasure now than they did in the past, knowledge-intensive work requires lots of reading. To understand how people make choices when they read, how they choose between printed and digital media, and what they find valuable in each, I interviewed a number of academics, scientists, engineers, and psychologists. They read for work, for leisure, and to their children. The interviewees also range widely across media, reading everything from blogs, technical journals, and scientific preprints to philosophical tracts and scholarly monographs, novels and poetry, alphabet books, and young-adult literature.

Like all of us, they live in a literary quantum world in which words can be either physical or digital. What surprised me was that *all* of them have constructed their own versions of Bohr's complementarity principle, which argued that an electron could appear to be either a particle or a wave, depending on the observer. For these readers, words can be bits or atoms, depending on the affordance of screen or paper.

The term *affordance* comes from Abigail Sellen and Richard Harper's *The Myth of the Paperless Office.* (Harper runs the Socio-Digital Systems group at Microsoft Research Cambridge and was the person who invited me to spend a sabbatical there.) In the course of examining why paper has survived in offices, laboratories, computer-equipped police vehicles, and even air traffic control centers, they discovered that people depended on the physical qualities of paper to work effectively, both on their own and with others. While computer makers treat paper as an unfortunate technological cul-de-sac and see its physical existence as evidence of human weakness, workplaces depend on paper's lightness, portability, flexibility, and malleability. Paper's affordances—the physical properties that allow it to be used by different people for different functions—are actually strengths.

The readers I talked to thought a lot about how the affordances of printed and digital media matched their reading practices, the nature of the material's content, the way they planned to read, and what they planned to do with what they read. For all, the Web is their source for news and for information that is highly changeable or that can be mined quickly for interesting nuggets and then forgotten. None of them defend print books on grounds of cultural authority or nostalgia. They choose print books because they're physically rugged and support serious, engaged reading.

As Elizabeth Dunn, an anthropologist at the University of Colorado–Boulder, put it, "If I can skim it, I read it on Kindle, but if I really have to know it, I need a printed page. Work I have to really concentrate on and know, anything I have to annotate, and poetry are all impossible to read on the Kindle. I cannot retain enough of what I read on Kindle to use it for things I will have to work to master." Stephen Herrod, the chief technology officer of VMWare, concurs. He'll take his Kindle on the road, but "I tend to print out deeper articles from the web if I know I'll need to think more about them."

Books and print are valuable when you need to read intensively and with few distractions. What's interesting is that reading books is often a very physical kind of reading. People talk about having to underline, annotate, and move between several books at once in order to engage in a kind of nonlinear, cross-textual reading. Novelist Nancy Etchemendy said, "If I'm going to use a book for reference or for mastering a difficult body of material, I find it helpful to annotate and bookmark. It is still impractical to do those things on e-readers." Microsoft Research scientist Ho John Lee says, "If I receive something important online, I print it out first, so I can spread it out, mark it up, and make notes. I never do any important sustained reading online." The stability of physical books is also useful for people with strong visual memories. Dunn remembers ideas by visualizing their physical location on the page. If she reads a book on a Kindle, she "can't remember a thing about a book beyond its general argument." Consequently, "I do almost all of my labor and thought-intensive reading in print form." For me, serious reading involves marking up, underlining, and annotating books; it's a martial art, and it requires the material engagement and support that paper can provide and that screens conspicuously lack. Fifty years ago, MIT professor and hypertext pioneer Vannevar Bush imagined we'd do this kind of intensive, interactive, relational reading on the memex, an electronic system he proposed in 1945. Today, people who really have to know their stuff still choose paper.

This need to support physical engagement with words explains why none of these readers buy digital books for their children. It's still an article of faith for them that babies need physical books that they can interact with (*Pat the Bunny* on an iPad would not be the same) and chew on. Reading from an actual book to young children who are sitting beside you or snuggled on your lap and settling in to fall asleep is a very physical and interactive kind of reading.

Forms of reading that are less intellectually intensive, make fewer demands on either media or reader, and depend less on affordances go digital more easily. Elizabeth Dunn's Kindle nicely illustrates how these choices work. It's stocked with fiction and academic articles she needs to "know about rather than really *know,*" and she pulls it out "any time I need to kill half an hour." E-readers are beloved by travelers who want to be able to carry lots of novels with minimum weight or who visit far-off and exotic locations (one person I interviewed took her Kindle to Antarctica). Nobody wants to be stuck reading the in-flight magazine on a trip to Dubai. The affordance of self-lighting that some e-readers have is useful: one engineer likes being able to read without turning on the lamp and disturbing his wife.

Finally, very targeted, opportunistic reading, the sort that you do when you're focused on finding a particular piece of information or need a quick overview of a new subject, is also almost entirely digital. When you can click from a citation of a legal opinion to the opinion itself, highlight a quote, and paste it into a memo that's due at the end of the day, there's little point in reaching for a printed volume.

For these serious readers, who've all read for years, who read a lot, and who need to do it well, print and digital media are not interchangeable, and the choice isn't made randomly. Each has its own advantages and supports *different* styles of reading.

Nearly everyone I interviewed had a dedicated e-reader, but no one saw the device as a replacement for books; rather, the device was used for lower-intensity reading. What's interesting is that this is not a distinction that e-reader manufacturers make—it's something that serious, thoughtful readers discover for themselves.

This sensitivity to affordances explains why many writers move between printed and digital versions of documents they're working on. Computers are great for writing quickly, while printed documents are good for seeing structure, measuring how well an argument flows, and getting a sense of a piece's overall balance and tone. For some writers, the act of marking up a physical document has subtle satisfactions: filling a manuscript with notes, annotations, corrections, and Post-its makes the work you've done visible. Physically marked-up documents can be easier for editors and coauthors to follow than electronically edited ones. A writer can quickly get an overall sense of how extensively a collaborator has changed the work and then determine how strongly he feels about making or reversing edits.

This affordance explains why some of my most tech-savvy friends are particularly enthusiastic about using groupware tools when they're in the same rooms as their collaborators. Programmers designing collaborative authoring systems usually imagine their clients being separated by oceans and time zones or doing work on shared documents at different times. This happens, but just as readers have discovered how printed and digital media support different kinds of reading, groups of coauthors have found that collaborative tools are most powerful not when the members are dispersed, but when they're together.

Again, it's all about the affordances. Writing with others tightens attention, as the presence of people with whom you're working and to whom you're accountable can keep you from being distracted. Additionally, a lot of the back-and-forth that defines coauthorship happens a lot faster. When I'm sitting with a coauthor and we're working on the same document, we can brainstorm a transition or talk about how to restructure a paragraph and then try it out immediately.

It radically lowers the cost of experimenting with changes, because they're more easily reversible. Historian of technology Ruth Schwartz Cowan recalls that when she and her husband coauthored a book in the late 1980s, they started out by writing on typewriters but soon ran into problems. "I would do a draft of the chapter, hand him the text, and he would cut and paste and muck it up. He'd hand it back, and I would be *livid,*" she recalls. "I couldn't put back things I thought were really good." When they started working on a word processor, though, "we instantly discovered that it kept us from being furious with each other because there was always an intact copy of what we'd done." They could look at an edit, talk it over, and try something else. "Cutting and pasting on paper was an emotional disaster" because of its permanence. On a computer, though, "it was completely different." Working on a word processor turned edits into suggestions to be negotiated.

Cowan's experience points to another critical affordance of digital, present collaboration: It lets you see body language, hear inflections in the conversation, gauge your coauthor's enthusiasm for a new idea or resistance to an edit. There's a lot that people communicate nonverbally that is easy to spot in person but that needs to actually be said when IM-ing or marking up a document remotely. Collaborating on something as cerebral and personal as a piece of writing is smoothest when you can see how your coauthor reacts to suggestions, and if you can see those reactions in real time, it's easier to avoid going down a path that will hobble a relationship. You can also learn a lot more from collaborators when they're sitting beside you. After all, in a good working relationship, you're doing more than just coproducing a document—you're building a connection, exchanging ideas, and learning about the crafts of writing and editing. This happens much more quickly with someone you're sitting beside. Online collaboration systems facilitate coauthorship at a distance, but they supercharge coauthorship in person.

You also need to take a more ecological view of your relationships with technologies and look for ways devices or media may be making *specific* tasks easier or faster but at the same time making your work and life harder.

A great example of the ironies of automation comes from the history of household appliances. In the century and a half since the Civil War—a period that saw industrialization, the invention of electric light and power, the arrival of the automobile and airplane, the growth of cities and suburbs—the American home has been mechanized and automated. The machines we live with may be smaller than those we see in factories, but vacuum cleaners, dishwashers, and washing machines changed housework as profoundly as the assembly line and electric motors changed manufacturing. Yet time studies of housework conducted decades apart showed virtually no change in the number of hours women spent keeping house: women in the 1970s spent as much time doing dishes, washing clothes, and cleaning house as their grandmothers had. Technology made housework easier, but it didn't make life easier.

Ruth Schwartz Cowan unraveled this paradox in her 1983 book *More Work for Mother.* Cowan is part of a generation of historians of technology who pushed the field to pay attention to users as well as inventors and entrepreneurs and to take everyday technologies seriously. This approach was unpopular at first. "None of my colleagues [at SUNY Stony Brook] wanted to talk to me," she recalls, speaking from her home in Glen Cove, a suburb of New York City. But it paid off massively in *More Work for Mother,* a book that, judging by the reaction people (particularly working mothers) had to it, could just as easily have been titled *You're Not Crazy, Housework Really Is Still Hard.*

At first, Cowan didn't believe the results of the time studies. After all, putting clothes in the washing machine *is* easier than filling a tub and using a washing board. "I really was not expecting what I found," she says. But what began as a straightforward narrative about innovation turned into an elegant cautionary tale of how technology makes work easier but then creates new work by changing *who* does the work and raising the standards to which it must be done.

Before it was automated, Cowan discovered, housework was a more gender-neutral activity. Wives were hands-on managers. Husbands and sons beat the rugs and wrangled heavy objects during spring cleaning, handled the horse and carriage, and helped with the shopping. Daughters worked with their mothers in order to learn to manage a house. All but the poorest households sent out their laundry to washerwomen and had domestic help at least a couple of times a month.

With automation, housework was *redefined* as women's work and, specifically, the mother's work. Further, the mother had to work alone; expensive appliances were paid for by firing the maid and doing the wash at home. Standards also went up. The massive spring cleaning involving the entire family was replaced by year-round daily vacuuming and dusting. Clothes were no longer worn for several days, or until they were sweaty and soiled, and then sent out to be laundered; now, at the end of the day, clothes went into the hamper, to be washed and folded by Mom. In other words, "There's no question that labor-saving devices save labor," Cowan says. "But they also *manufacture* labor. The washing machine and the dryer don't save time if you end up doing more washing."

Cowan had discovered housework's equivalent of the Jevons paradox. In 1865, English economist William Stanley Jevons observed that demand for coal was not decreasing with technological innovation and better energy efficiency. Rather, factory and mine operators with access to new, more efficient coal-fired engines increased production or installed engines in parts of their facilities where previously the cost of doing so would have been prohibitive. "It is wholly a confusion of ideas to suppose that the economical use of fuel is equivalent to a diminished consumption," he declared. "The very contrary is the truth." Increased efficiency triggered *increased* use of technology, which led to an overall increase in energy consumption.

Economists argue about how pervasive the Jevons paradox is, but as Cowan notes, labor-saving technologies often seem to lead people to "choose to do things that consume more labor, more time, more energy." And new technologies are never deployed in a vacuum; the worlds they're introduced into are always changing. Women's work became more time-consuming when families moved from cities to suburbs and women had to take on new duties as chauffeurs and carters; they delivered the kids to school and the husbands to work and drove to supermarkets (which were now too far from home for the children to go to).

Technologies operate in contexts and often are parts of larger technical or productive systems; as a result, improving one part of a system affects others, sometimes negatively. Antilock brakes, which work better than regular brakes in hazardous conditions, ultimately don't make driving safer; antilock brakes inspire drivers to go faster, trusting that their better brakes will keep them out of accidents. A similar dynamic is at work in American football. Even though padding and helmets have become more sophisticated over the decades, injury rates haven't fallen, because players have become larger and more powerful, and the game has grown more physically demanding.

The case of housework illustrates how the term *work* means somewhat different things depending on whether it's describing technological or human activity. For machines, *work* is narrowly defined. Machines are designed to wash the clothes, collect dirt from the floor, clean the dishes, or do some other specific thing, and they work when they do them. For people, though, work is rarely so simple, especially when new technologies raise the possibility of working in new ways and when they change performance standards. After the introduction of washing machines, what had been weekly laundry became a daily event. Thanks to cell phones and e-mail, lawyers are expected to be there for clients day and night, and bosses take for granted that their employees will be available to work weekends; the fact that someone *can* be accessible means she *must* be accessible. We're told that new devices will save us time, and then we wonder why we have no time.

Our interactions with digital devices and media offer opportunities for self-improvement as well as self-experimentation. An algorithm developed by Finnish entrepreneur Jarno Koponen taught me that.

A smart, slight man with graduate degrees in intellectual history and design, Jarno could be the keyboardist for a cerebral Scandinavian post-punk band; instead, he's the founder of a startup called Futureful. We meet at Peet's Coffee in downtown Palo Alto; Steve Jobs has just died, and the flagship Apple store across the street is covered in Post-its, flowers, and pictures from mourners and fans. It is a great reminder of how technology can touch our hearts and how much we can warm to a universe of zeros and ones.

One enduring criticism of the Web and social media is that they both encourage narrower, not more expansive, reading. Lots of studies have concluded that our social graphs and reading habits replicate our real-world prejudices and political views: there's remarkably little overlap in the readerships of left- and right-wing political blogs, for example. We have enough trouble keeping up with our friends and favorite Web sites to really explore the world of information; confronted with overwhelming volume, we retreat to the familiar. Futureful is trying to reintroduce serendipity in people's online reading, to help users discover articles, writers, and Web sites that they're not familiar with but will like. In order to do this, the organization has to know who a person is. Based on the user's online activities, Futureful builds a model of what he's interested in. In effect, it shows him what he looks like to the rest of the Web.

At Peet's, I take my laptop and log in on Jarno's demo; Futureful dives into my accounts, and the algorithm starts to analyze my blog posts, tweets, and likes. As we're waiting for results, someone at the next table asks, "Did you say you were with Futureful?" Jarno and his company are based in Helsinki, and it's his first time in California, but a random individual here has heard of this little five-person startup on the other side of the world. Typical Silicon Valley.

The results come up on a typically sleek Finnish interface. I've always assumed that my Facebook and Twitter pages reflect my true self. But when I look at the Futureful algorithm's profile of my interests—its sense of who I am, its snapshot of what I look like online—I'm puzzled at first, then actually alarmed.

The person Futureful thinks I am is very interested in politics; most of its recommendations are from partisan American Web sites or European news sources. (To the program's credit, most of them are new to me; the system *is* doing what it's supposed to.) According to Futureful, I'm also very cynical. It thinks I like to read about corruption, scandals, and disasters caused by shortsightedness and greed. There's nothing about history, design, computer science, or futures. Nothing about

Buddhism or religion. Nothing about science. This person is an observer of the follies and stupidity of mankind, a cybernetic H. L. Mencken.

If I were talking to this person at a party, I'd concoct an excuse to get away from him.

Of course, it *is* me; no one has hacked my accounts. But my digital shadow seems cast from a very odd angle, and the result is—I hope—quite distorted. What's going on?

Jarno explains in greater detail how Futureful works. The algorithm accesses a user's Twitter, Facebook, and LinkedIn accounts. Those services have encouraged third-party developers as a way of building user loyalty, and they have created application programming interfaces that are easy for programmers to use. It makes a lot more sense for startups to focus on these big three—which together have over a billion accounts—than on smaller, niche services like Zotero and Delicious.

For me, though, this means there are large swaths of my online life that Futureful can't access yet. If it could analyze my Delicious account, for example, and see the thousands of scholarly articles and books I'd tagged, it would get a very different picture of me.

But as Jarno talks, it dawns on me that, for better or worse, people are much more likely to use Facebook and Twitter to build their picture of who I am than they are to read through my Delicious bookmarks. The system has limits, but they neatly replicate the Web's own preferences.

So why do I use them that way? It's much easier to share something on Facebook or Twitter than on Delicious; you can do it almost without thinking.

And there it is—almost without thinking.

I tend to fire up Facebook and Twitter when I'm taking a break from more serious work, when I'm not really thinking that much about what I'm doing, and when I'm at my least mindful.

Is it possible for me to engage with Twitter and social media in ways that don't accentuate the darker, snarkier sides of my personality but instead make me appear more like the person I want to be, both online and in real life? Could I actually be contemplative using Twitter and Facebook?

Marguerite Manteau-Rao puts it another way: "If the Buddha was alive," she asks, "would he use Facebook or blog? I think he would." I've sought out Marguerite because she manages a substantial online presence (she has five thousand followers on Twitter) and writes extensively about mindful use of social media. I admit I feel a bit uncomfortable in her company at first. I'm tempted to dislike people who have great poise—they make me feel trollish and uncultured—but it's impossible to resent Marguerite's meditation-sharpened elegance. She approaches her own contemplative practice with the seriousness of an athlete. She's also a leader in efforts to use mindfulness to improve dementia care and help caregivers be more able and compassionate with patients, spouses, or parents whose minds are slipping away. You try being annoyed with someone who does that.

The Buddha would blog, she continues, because it's "a great way to reach the *sangha*." Her French accent gives the word *sangha* (which is Pali for "community of the faithful") an unexpected European softness. But, she continues, if approached the right way, social media also offers opportunities to practice mindfulness.

First and foremost, members of the digital *sangha* express themselves with care. "Twitter is an opportunity to practice right speech," Marguerite declares. *Right speech* means, at the very least, that you're not cruel or chatty. Mockery and triviality are right out. Less is often more.

Others practice right speech by tweeting lines from the Pali Canon or chapters from the Bible, 140 characters at a time. Elizabeth Drescher, a professor of theology at Santa Clara University, explains that this kind of sharing—copying out, reading, and reflecting on short passages—is an updated version of very ancient practices. "Much of the Bible is *way* tweetable," she tells me, because the "Scriptures are strings of memorable memes," meant to be easily recited, meditated over, and discussed.

Tweeting mindfully means knowing your intentions; knowing why you're online right now and asking yourself if you're on for the right reasons. (As she speaks, I notice that Marguerite sometimes talks about "using" Twitter and sometimes about "being" on it. Interesting.) As a practical matter, this means that if you read something and your first impulse is to post a sarcastic comment or to blather on, stop and consider why this is the case. Marguerite admits she sometimes uses Twitter as a diversion at the end of an intellectually challenging day. That's not inherently bad, she adds, but it's always important to be aware of your own state and adjust your own behavior accordingly. As a reader, you shouldn't be afraid to unfollow people when your interests or lives diverge and to stay offline when it's more important to be present elsewhere.

You should always be mindful that you're interacting with *people,* not just processing text. The technology and words are just means. You're reading or following or retweeting things written by real people, and the fact that you're interacting through technical intermediaries should not distract you from their humanity. This focuses your attention on the quality of your connections, not the quantity. "If I'm going on Twitter as a Christian," Elizabeth Drescher says, "my aim is to see Christ in everyone" and to encourage others to see Him in themselves. Drescher is the author of *Tweet If You Love Jesus,* and she's been explaining the Web to mainline Protestant pastors who have been reluctant to take their churches online. "Digital ministry isn't interested in evangelizing," she tells me. "I have no interest in *marketing* the church. Zip.

Zero." Rather, she believes that pastors and people of faith should think of social media as a way to be "a spiritual presence in the places where people *are,* to deepen relationships, so that in real embodied ways, lives are transformed."

The digital *sangha* lives first and tweets later. This means you shouldn't feel the need to provide a play-by-play of everything you're doing, even if it's novel or interesting. Indeed, there are pleasures and insights that can emerge from crafting a story out of episodes in your life, and distance can give clarity and meaning to events; something that in the moment seems catastrophic can have wonderful consequences, while a victory can lay the groundwork for a later setback. You risk making less sense of your own life if you reduce it to a running set of descriptions. Having experiences worth writing about and thinking enough about them to make the writing worthwhile is more important than saying a lot quickly. Experience now, share later, and give yourself time to make sense of what you've done. Members of the digital *sangha* are deliberate, not reactive; they write when they have something to say, not when someone else speaks. Mindfulness writers advocate saving tweeting for specific times (twice a day, say, in the late morning and late evening), moods (when you need a break), or milestones (after you've checked the next task off your list). That keeps the feed in its place and prevents you from spending unnecessary time online.

These rules show that the nasty, argumentative, cruel tone of many online exchanges is not in any sense inevitable. Plenty of people feel that they can be rude when they're anonymous, or they find that the computer makes it easy for them to dehumanize others or that it's fun to be part of a mob. It might even seem that there's something in the nature of the Web that makes people act in unsocial or amoral ways. But we can engage with social media mindfully; even though some people act like troglodytes online, we can choose to behave very differently.

Equipped with these rules—engage with care; be mindful about my intentions; remember the people on the other side of the screen; focus on quality, not quantity; live first, tweet later; and be deliberate—I set out to see if I could make my tweeting sound more like myself.

The first thing I notice after a few weeks of following these rules—or at least trying to follow them—is that I'm on social media a lot less often, but a lot more purposefully. I cut way down on the reposting and retweeting. I don't eliminate it entirely, but if 17,000 people have already liked a video, do I really need to make it 17,001? I start treating social media as an opportunity to focus on what I'm doing, determine whether it matters, and decide whether it matters enough to share. More often than not, I conclude that my friends don't need to know what I'm up to.

This has an effect on the quality and tone of my posts. When I'm not thinking about them, my Twitter stream and Facebook page can still be pretty mindless. But when I really pay attention to what I'm doing, my Twitter stream looks more like a Renaissance commonplace book (a personal journal used to record favorite passages, literary excerpts, and comments), heavier on quotes and references to interesting articles and to other people's work. On Facebook, days can pass when all I do is wish a friend happy birthday. (I love the birthday-reminders feature.)

I also don't play so much to the crowd. In my more distracted periods, I'd check my account several times a day to see if anyone had retweeted or liked something of mine. Once I start doing social media more mindfully, I don't feel the same pressure to post constantly or be entertaining. After a few weeks, I realize that I no longer have any idea how many followers and friends I have. Social media can provide plenty of positive reinforcement, and there's a lot of satisfaction in having a growing number of followers. But numbers are no longer nearly so rewarding. Contact with people is.

Social media is like the stream you can't step in twice; for all our worries about old pictures or comments haunting us, it can be surprisingly hard to visit a specific moment in our social past. When you give up trying to follow all your friends all the time, you need to accept that you're just going to miss some fascinating things.

As I accept these truths about social media, I find myself also accepting its ephemeralities. It's constantly changing, and I realize I'll never really be able to keep up with it. At best, staying on top of my Twitter and Facebook feeds is like trying to stay involved in a dozen fascinating conversations at a party. As stimulating as that can be, it's too much to keep track of if I ever want to have my own thoughts. Embracing the ephemerality makes it easier to let go of my efforts to keep up. It also means coming to terms with the fact that most of what I write will eventually become inaccessible, and my own thinking will change. But that can be a feature, not a bug. "Don't get attached to opinions," Marguerite advises. "Opinions are not very interesting anyway, and to get attached to them and protect them—that's *really* uninteresting." The disappearance of old ideas clears the way for new, better ones.

Becoming more mindful about how you use technologies can make you more attuned to the unexpected abilities that you can cultivate when using technologies and the ways in which information technologies can extend your mind. You're more likely to be aware of how the subtle affordances of different media can support or subvert your work. You'll also be more aware of how they can help you develop new skills.

For example, I've found that geotagging pictures makes it easier for me to remember my travels and gives me a clearer vision of my world.

Here's how.

For years I've used the photo-sharing site Flickr. One of my favorite features is its mapper. To associate a picture with a place, you put a digital pin in an online map, much as you would in a real map. Flickr and Yahoo! Maps got together to provide the service in 2006, and since then I've become a slightly fanatical geotagger. It started out as pure geekdom. I'd written about the future of geolocation services, so it seemed a good chance to play with a future I had described. (As a general rule, I don't tag pictures of family or friends, out of privacy concerns, and there's no cognitive benefit to tagging familiar places. The newer and more exotic a place is, the farther it is from home, the more likely I am to geotag it.)

When I travel to a new place, I like to walk. I want to know enough to stay out of bad neighborhoods, to find interesting ones, and to be aware of significant landmarks. I don't want to miss the big attractions, but I also treasure the experience of turning a corner and finding a perfect little café and pastry shop or the brilliant bookstore that's not in any of the guidebooks. (How many travelers define themselves as people who want to escape the boundaries of the guidebooks?) As a result, I absolutely love cities that reward walking. In London, you can't go three blocks without coming upon something grand and historic, a charming little square, or an interesting piece of street life. To paraphrase Samuel Johnson, when you're tired of walking around London, you're tired of life. Singapore has a tropical city's mix of overflowing gardens and pools, brilliant architecture from three centuries, and amazing food. Budapest is a wonderful Old European city of twisty streets and grand boulevards, the magnificent Danube, and faded (but rapidly being renovated) buildings and apartments, with great coffee on every block.

So I like to wander. Once I'm back at the hotel, I like to reconstruct my wanderings and figure out where I've been. I used to do this on paper maps, tracing my route with a highlighter. This required remembering street names, knowing how many blocks I walked before I turned left, estimating how far I'd gone on the boulevard or embankment before I stopped to take those pictures. Given that I often walk at night—since clients almost literally own my days—all this was tough. The map I put the information on was frequently in an unfamiliar language, which didn't make things any easier. Then, as often as not, I'd forget the map when I left.

Flickr's mapping program made all of this much simpler; I could reconstruct my route more quickly and look back at it later. But a couple other things turned me into a digital-mapping fanatic.

Like many digital maps, the Flickr map offers regular map views (ordinary road maps with labeled streets, rivers, train lines, and so on), satellite views (aerial photos), and hybrid views (a combination, with aerial photographs overlaid atop the street map). Satellite mode let me establish much more precisely just where I had been, what the photograph showed, and where it should go on the map. Without it, I can put pictures on the right block; with it, I can get to within a few feet. But it required decoding aerial photographs and learning how to relate that information to my own experience.

Unless you've worked for the CIA or had a particularly sadistic geography teacher, you've never had the chance to do this. Connecting one's ground-eye view of a landmark or city grid with an aerial view isn't that hard, but it does need to be learned. When it goes well, it becomes a game of imagining how what I've seen would look from space. London's Trafalgar Square becomes a long shadow (Nelson's Column) with a few shapes (the lions around it, the fountains nearby); Leicester Square becomes trees and park paths bordered by the blocky shapes of theaters. Sometimes I realize how big something actually is ("Boy, Suntec City really is *huge*"). When I'm trying to find someplace I've reached by taxi or subway, if I know the shape of the building and have a pretty good sense of the buildings around it, I can find it on a satellite map.

Pinning pictures on a Flickr map combines three different kinds of knowledge. It draws on the physical memory of travel—your feel for where you went, how far you walked. It utilizes visual memory, reaching across your extended mind of biologically stored memories and into the domain of silicon. It then encases your physical and visual knowledge and memory in a formal system: the logic of the map. By linking these all together, you connect your personal, street-level view of a place with a formal, high-level view. They're your memories, organized. And in the course of organizing your memories, you build your knowledge of the place and how it's laid out.

You could argue that my learning how to look at satellite pictures isn't useful when I have to use an old-fashioned street map. Maybe so. One of the things we have to be aware of is how these new skills sometimes come at the expense of older ones, and we have to make conscious choices about whether we want to abandon them or not.

One great example of the complex choices information technologies force you to make in creative work is in architectural education and practice. Technology has made it possible for architects to explore new geometric forms, simulate a building's resource consumption, and walk clients through virtual models of draft plans. It's also killed drawing. And almost everyone in the profession is okay with that.

For centuries, drawing was fundamental to architecture. Artistic skills set the architect apart from the mason, the carpenter, and the mere tradesman. Drawings and blueprints were media in which architects communicated with builders and clients. And, most important, drawing was the medium in which architects thought. Learning to draw taught you how to observe the world and how to express yourself. Drawing designs and plans was very labor-intensive; architectural firms hired armies of draftsmen to create plans and elevations, and changing plans was costly and time-consuming. Computer-aided design (CAD) greatly reduced the cost of producing blueprints and architectural drawings, but in the past twenty years, it's done much more.

Some architects have exploited CAD to create forms they could not have drawn with pen, paper, T square, and compass. Most notably, Frank Gehry's buildings, with their swooping, curving surfaces, would have been impossible to design and build without CATIA, a design program that had its origins in the aerospace industry. Simulations allow architects to forecast a building's energy use (a big issue in today's world), to model traffic flow in large projects like airports and shopping malls, and to see how buildings can be fortified to withstand terrorist attacks and earthquakes. They let architects (and, just as critically, clients) see how a building will look with different materials. CAD files can be quickly shared with subcontractors, engineers, and construction companies, which facilitates scheduling and budgeting and makes it easier to deal with last-minute design changes, budget shortfalls, or delays. (Sharing has unexpected benefits; futurist and urban planner Anthony Townsend says that after the attacks on the World Trade Center in 2001, "architecture firms in the WTC were able to recover their information losses by getting CAD files back from their clients." CAD makes clients into redundant archives.)

In other words, architects no longer just make use of computers; they think with them. The computer network is the nervous system of a firm, the medium through which an architect communicates with clients and builders and governments, but it's also an extension of the architect's mind.

Since architectural practice has gone virtual, drawing has all but disappeared from architectural education—and this has had effects on the way architects think. In the 1990s, architecture schools started dropping drawing from their curricula. Students who could draw but who hadn't mastered CAD, deans fretted, wouldn't find jobs in the always-competitive marketplace, and students could work faster with computers.

But digitization and convenience have made architectural education less rigorous, according to Witold Rybczynski, an architecture professor at the University of Pennsylvania. Architectural drafting used to be a key skill that architects had to learn as students and develop throughout their professional lives. Drawing gave the best students a more intuitive, well-developed sense of proportion, a sharper eye, and a better imagination, and it let them think more effectively about technical issues. The physicality of drawing—the constant interaction between pencil, paper, and imagination—and drawing's slow pace created a chance for architects to be more contemplative and engaged, even for them to make mistakes that inspired new solutions.

With computers, by contrast, it's almost trivially easy to generate a large number of prospective designs, make changes, and create smooth and polished-looking drawings. This can keep students from thinking seriously about basic architectural questions. "The fierce productivity of the computer carries a price," Rybczynski writes, "more time at the keyboard, less time thinking." Architectural historian David Brownlee, a colleague of Rybczynski's at Penn, complains that CAD "makes all student work look the same." Of course, architecture has its fashions, just like any art, but drawing gave students' work a measure of distinctiveness. Today, "the sameness is driven by the technology." The exact, crisp look of CAD drawings also leaves less room for experimentation or play; because drawings never have the roughness of sketches, ideas look finished before they're fully thought out.

The problem continues after the students graduate. As architect Renzo Piano explains, when architects work on extremely complex buildings—particularly airports, city halls, and other high-profile, signature buildings that clients want to use to both make a statement and, well, actually use—"you need a computer to optimize everything—the structure, the form." CAD systems are invaluable for helping architects keep track of details, anticipate how changes in one element will affect others (how increasing the size of windows will affect the demand for air conditioning, for example), and model how the building will look under different conditions (something especially valuable for clients who may not have a designer's strong visual imagination).

But working with these complex, fast tools narrows the window of opportunity for architects to think deeply, to consider the site and program, to discern what the clients will really accept, to treat ideas as unfinished and tentative. As Piano puts it, with today's systems, "you may find yourself in the position where you feel like you're pushing buttons and able to build everything. But," he continues, "architecture is about thinking. It's about slowness in some way. You need time. The bad thing about computers is that they make everything run very fast, so fast that you [think you] can have a baby in nine weeks instead of nine months. But you still need nine months, not nine weeks, to make a baby." As Chicago architect William Huchting tells me, "Architecture is first and foremost about *thinking*... and drawing is a [more] productive way of thinking" than CAD.

The field is still looking for "the right balance between the digital tools and the physical world," Chris Luebkeman says. He's a futurist with Ove Arup and Partners, a global engineering firm that pioneered the use of computers in architecture. (In the early 1960s, they used an IBM mainframe to help design the shells on Jørn Utzon's fantastic Sydney Opera House.) Today, Luebkeman says, design tools are "amazingly wonderful, *and* absolutely horrendous." On the creative side, CAD lets architects "see airflow, heat flows, and truly understand the performance of your place and space." Yet it also teaches some students that "if they can draw on the computer, that makes it real. You can make a well-rendered but terrible space look good, and create architecture that looks good on a screen but is inhumane, has terrible construction details, and so on."

CAD has also had an impact on the culture of architectural and engineering practice, on the way younger engineers learn from their bosses and mentors, and that impact hasn't been very positive. "Twelve years ago we still had big tables and big rolls of paper" in the Arup offices, Luebkeman says, and "people were still doing plan checking, and the senior guys still literally looked over the shoulders of the junior people." Today, though, "It's not culturally appropriate to look at someone else's screen, because it's their private domain. We've lost some of the tacit transfer of knowledge that happened at those tables. We've observed that, and we try to counter it with project reviews in semipublic spaces. We're trying to counter the loss of the watercooler effect"—the process of informally sharing knowledge and teaching that happens in good workplaces—"but it remains a challenge for us, and a challenge to the profession as a whole."

Still, nobody can imagine—and nobody proposes—abandoning computers for vellum and ink. Creating contemporary architecture without computers and the Internet would be impossible. Digital tools are far too deeply integrated into the everyday life of architecture. Piano's practice is organized around the kinds of vast projects whose built forms are like the tips of icebergs, supported by vast digital infrastructures. What architects are calling for instead is an effort to recover skills that were lost in the move from drawing to digitization.

When I tracked my own e-mail use, I was surprised at the results. When I counted up the number of times I checked it at my computer, while standing in line at the bank, and when waiting for my kids (or for the light to change, I hate to admit), I saw I spent easily an hour a day waiting for mail to download and doing account management, and I spent twice that replying to messages. In the moment, it *felt* like work, but it didn't have any lasting value. Messages from my e-mail-happy family aside, on most days, my in-box would have at most a handful of messages that needed my immediate attention. The others were reminders, updates, ads, replies-to-all, or junk.

So I started experimenting with my e-mail habits, trying different things, seeing what worked better.

I unsubscribed from 99 percent of my lists and newsletters, and I switched off all alerts on my iPhone while I was at it. I don't want to give a message the power to inflate its own urgency or to impinge on my attention before I'm ready. My working life doesn't have enough emergencies for me to need instant updates. The great computer scientist Donald Knuth gave up e-mail in 1990, declaring that while e-mail is "a wonderful thing for people whose role in life is to be on top of things... my role is to be on the bottom of things," to do fundamental research that required "long hours of studying and uninterruptible concentration." Now when I check my e-mail, I don't look at the screen immediately; I'll hit Refresh, put down my iPhone or look away from my laptop's screen, and focus on something else while it works in the background. Deliberately ignoring the program rather than watching it connect to the Internet doesn't make it work any faster, and it may not make me more productive, but every time I do it, it's a little declaration that my attention will go where I direct it to go and will not get caught in a Web. It's a way of stating that I'm in charge.

I've tried to apply the social-media principles Marguerite Manteau-Rao outlined to my e-mail too. Every now and then, think about these rules when you're composing a message. Is this message really necessary? I'll ask. This person already gets a lot of e-mail; will this message be welcome? Would it be better if I picked up the phone? Could I avoid six hours of back-and-forth and a ten-message chain by going down the hall and talking to my colleague? If I'm part of a round of messages with several people, would it be better to reply to every new message or to weigh in at day's end with a less in-the-moment response? The point is not to put myself out of touch but to use the technology in ways that serve everyone better, ways that reduce internal and external distractions and let me give people as much attention as they deserve.

I've set times when I deal with mail, and I avoid it the rest of the day. Because I feel good responding to mail that matters and dislike having nothing new, checking less often eliminates an emotionally negative experience. I've tried handling e-mail on one device only, either my laptop or my iPad. (I go back and forth on this one. If I'm working on a book, segregating the action of checking e-mail can be helpful, but it's not a big deal.)

I keep experimenting with e-mail, and I have no doubt that as my attention span changes, as new programs become available, and as standards and norms and friends change, I'll need to keep experimenting. But equipped with a few basic tools and a habit of asking whether what I'm doing is improving and expanding my extended mind, I think I'll discover ways to find new balances, to reach a state of detached, calm engagement that is essential if I'm to be a thoughtful correspondent and have a well-working extended mind.

"We need to be mindful about what we're doing with technology," Ruth Schwartz Cowan agrees. For her, the history of technology is what made her more mindful and gave her a deeper awareness of the complexity and elastic nature of humanity's relationships with tools. That awareness changed the way she lived. Writing *More Work for Mother* and thinking about "the history of mundane technologies, the technologies of everyday life," she says, "helped me figure out that my ultimate goal as a housewife was to get everybody out of the house, to allow me to do something else, let my children be healthy and go to school, so that they could grow up to get out of the house." She and her husband talked a lot about just how to make that work, and if some habit "didn't help us do that, we stopped. And it changed my whole routine. Yes, everybody had to get fed, but did it have to be a gourmet meal? No! Did everybody have to sit at the table together and do the things that we should do as a family? Absolutely."

Recently, Cowan pushed back against e-mail creep. "I used to be a person who would check my e-mail once or twice a day," she says, but when she left SUNY Stony Brook for the University of Pennsylvania and started commuting between Long Island and Philadelphia, "like everyone else, I got sucked into checking my e-mail every hour or so on the train. At one point I stopped and realized that it was better to read, or even to sleep, than to check my e-mail." Cowan isn't talking about destroying the machine or creating a perfect world but about learning to live more thoughtfully with the world as it is. You need to be mindful "about what you are doing with the technology at hand," she says. "We need to focus on our ultimate goals and make the technology work for us, rather than have the technology set the goal. That's what mindfulness is. You have to get in the habit of thinking about the tool you're using, and whether it's getting you to the goal. If one brush isn't doing the job on the canvas, you switch to another brush. You're the craftsperson, you're the artist, you have a vision, and if the tool isn't working, get another one."

Learn how to focus on your ultimate goals, be mindful about the technology at hand, and switch tools when they don't work—this is how you enhance your extended mind, become a better presence and person, become a craftsperson of your extended mind. The right technology used the right way can help, but it can't make you mindful. The blogging monks teach us that contemplation is a skill that people can learn, refine, and practice even in an always-on world. They also teach us that focus isn't what's left over when we remove e-mail alerts, pop-ups, LOLcat videos, telemarketers, and links to the latest video of dogs dressed like Queen Elizabeth. Concentration isn't waiting to spring up, jack-in-the-box-like, when the weight of other stuff is removed. Concentration is an active, skilled engagement with a purposefully narrowed piece of the world. Design may invite or discourage mindfulness, but you have to make the choice to develop and use those skills. It's up to you to use technologies mindfully, and it's within your ability to do so.

Once you know how to use technologies more mindfully to take control of your extended mind, be prepared to fail. Because you will. But don't be discouraged. Instead, turn that failure into a chance to rest productively.

SIX

REFOCUS

The next time you're online and notice yourself losing focus, go to the Web site Do Nothing for Two Minutes (http://www.donothingfor2minutes.com/). The site features a picture of a beach at sunset, the calming sound of waves, and a timer that counts down from two minutes. *Just relax and listen to the waves,* a text in the center of the screen says. *Don't touch your mouse or keyboard.* If you do touch your keyboard, the word FAIL appears in a startlingly large red box, which is not at all relaxing. Some elements are still a little bit rough around the edges. One friend said that "two minutes of sitting still listening to the waves was quite relaxing," but the effect was ruined when she noticed that the horizon wasn't quite level. (You can stop doing graphic design, but you can't stop being a graphic designer.) Still, it's a brilliantly simple concept, an interesting attempt to create a technology that helps you refocus and recenter your attention.

It can be surprisingly effective. Rebecca Krinke, a landscape architecture professor at the University of Minnesota, uses the site in her class about experiments in design processes. When she feels the need to break them out of their usual habits of thinking or encourage them to reflect on their own processes, she has them all watch it together. "We noticed a shift" in the mood of the room, she said.

Most technologies try to capture and keep your attention, to direct it somewhere. Do Nothing for Two Minutes encourages you to pause, sit, and recenter yourself. It doesn't try to prevent your mind from wandering. It works with your mind's need to take a break. At the same time, it's not too engaging; there are no puppies sliding across kitchen floors, no pop-up windows. It suggests that we can use technologies that normally do a brilliant job of distracting us to help us refocus.

According to the map on the monitor in front of me, we're just south of the tip of Greenland. I'm six hours into a ten-hour flight from San Francisco to London. The rest of the cabin is a dark, vaguely perceptible space outside this pool of light. My iPod is plugged into a pair of noise-canceling headphones, which reinforces my sense of isolation and dampens the sound of the engines and the wind howling past at five hundred miles an hour. Most of my fellow passengers are asleep; a few are reading (this season, every international traveler is required to carry one of the books in Stieg Larsson's trilogy) or watching movies. Me, I've been working, and I will do so until sleep finally catches up with me. By the time I get through passport control, I'll be a zombie; I'll get on the bus to Cambridge and be asleep before we leave Heathrow. But it's worth it. I do some of my best thinking on planes.

For the past few years, I've spent a lot of time on planes, flying to pitch consulting services, meet with clients, deliver forecasts, or run strategy workshops. Travel has become a focused, businesslike affair. On business trips, as soon as I land, I often take a taxi to a conference center or a client's headquarters and get straight to work. When I fly to Europe, I normally need only a couple of days to help a government agency or corporation see the black swans waiting to hatch, and before the jet lag has time to wear off, I'm on a flight home.

You would think that a futurist would be a master of time management, but I'm always a deadline out of phase and a project behind. So work crept onboard with me, and I became one of those harried-looking people hunched over PowerPoint presentations and meeting agendas.

Slowly, though, I realized I could actually think well on planes. Cut off from the office, nearly seven miles in the air, free from distractions, and facing a nonnegotiable deadline of twelve hours to *get it done,* I had just the right blend of pressure and freedom necessary to work hard and fast. The length of a flight from San Francisco to London or from San Francisco to Frankfurt is fortunate: ten hours is enough time to rewrite a talk but not enough time to overthink it.

I feel the distractible, jittery parts of my mind quiet down, and the parts that concentrate take over. As I work, I shift from a state in which I'm wrestling with problems to a state in which I'm observing the problems solve themselves. There are times when I look over a paragraph I've just written and wonder, Where did *that* come from?

It's partly because my world is reduced to a few objects, all within easy reach. I've learned how to turn my tray table into a mobile office. Maneuvering in the tight space (business class is what I walk through on the way to my seat, alas) requires care and deliberation—everything has to be accessible, but nothing can get in the way of anything else. This evening, it's crowded with a travel mug, notebook and fountain pen, a book, and the latest draft of an article I'm working on, all illuminated by the overhead seat lights. I'm not facing a hard deadline on this flight, but I'm still too excited to do anything but work. I'm starting my fellowship at Microsoft Research Cambridge, and I've given myself a big challenge for the next several months: to figure out how computers can be designed to help people think deeply and not be distracted. Months ago, while discussing potential research projects with the lab, the phrase *contemplative computing* popped into my head; now I'm going to find out what it really means.

I'm a historian of science by training and was last in Cambridge twenty years ago when I was working on my dissertation. That time, I barely left the library; this time, I plan to make better use of the place. I've brought along two books as my guides: James Watson's *The Double Helix,* his account of how he and Francis Crick discovered the structure of DNA, and a biography of Charles Darwin, the father of the theory of evolution by natural selection. Watson was probably the most important American visitor Cambridge will ever see: in a little over a year, between November 1951 and February 1953, he and Crick beat several better-funded, more famous groups in the race to solve one of the great scientific puzzles of the twentieth century. I can't match the importance of their work, but I figure Watson's as good a guide as any for an American going to Cambridge with ambitious intellectual plans.

I've read both books many times but never as travel guides or how-to manuals; each might as well be titled *How to Be a Genius in This Place.* Now, as I leaf through *The Double Helix,* I'm struck by something. Watson and Crick come across as extremely ambitious and driven, but there's a lot of walking around in the book. On Watson's first afternoon there, his adviser John Kendrew gave him a tour of the colleges. He and Crick went walking after they had lunch at the Eagle, a famous pub founded in the 1500s, just up the block from the Cavendish Laboratory where they worked (I'll bring my copy of *The Double Helix* to several dinners there). He went for long hikes during trips to the Continent. After he and Crick solved the structure of DNA, Watson described walking "toward the Clare Bridge, staring up at the gothic pinnacles of the King's College Chapel that stood out sharply against the spring sky" as he reflected on his and Crick's accomplishment and "thinking that much of our success was due to the long uneventful periods when we walked among the colleges or unobtrusively read the new books that came into Heffer's Bookstore."

Charles Darwin became one of the most important scientists of the past five hundred years, yet when he arrived at Cambridge in 1829, his potential for greatness was hardly suspected. His father had sent him to the University of Edinburgh to study medicine, but young Charles hated the sight of blood and was distracted by natural history. Now, to all appearances, he was just another well-bred but mediocre student bound for a quiet clergyman's life. Instead, he discovered a passion and talent for science. His energy greatly impressed botany professor John Henslow. Other professors referred to Darwin as "the man who walks with Henslow," but they were probably happy to see him go. He wilted in the classroom. It was in the field, collecting and exploring, that he flourished. In 1831, Henslow was asked by the Royal Navy to recommend someone to serve as assistant naturalist aboard the HMS *Beagle,* a survey ship that would be mapping the coastline of South America. He suggested Darwin.

Darwin's five and a half years on the HMS *Beagle* observing the natural history of South America and the Pacific, collecting rare specimens, and sending back accounts of his travels established him as one of the leading naturalists of his day. After his return in 1836, he spent the next six years at the center of London's vibrant scientific world. When he moved to a quiet country house, he constructed a walking path on his property and could be found there at some point every day, turning about the grounds while thinking. Darwin spent his most fruitful years in motion.

These days we think of productivity or innovation as something that can be gotten from lifehacks, caffeine, or even medication; in a 2008 survey conducted by *Nature,* a fifth of respondents confessed to using Adderall or Provigil to boost their concentration and allow them to work longer. Yet two of history's most important scientists solved problems by *walking*. What was it about walking that helped them?

For nearly forty years, Charles Darwin took long daily walks on the Sandwalk, a quarter-mile path that began in the backyard of Down House, the Darwin family home. Charles and his wife, Emma, moved there from London in the summer of 1842 to start a family and escape the distractions of the city. Down House was a former parsonage set on three acres of gardens, and it had an additional fifteen-acre field divided into four meadows. Darwin's move to Down House is sometimes described as the act of a recluse. Some writers have contrasted his strenuous years aboard the *Beagle* and the busy whirl of London's scientific scene with his life in the quiet Kent countryside and concluded that Darwin was removing himself from society, even fleeing his own theory of evolution. The reality is far more interesting, as Open University professor James Moore explains to me.

Moore has been visiting Down House and the Sandwalk for thirty years and is as knowledgeable about Darwin's world, from the grand sweep of his ideas to the small scale of his domestic life, as anyone alive today. He sees Down House as equal parts home, sanctuary, laboratory, and fortress. Ultimately, he argues, it was as important a space as the *Beagle* in shaping Darwin's life and thought.

Both Charles and Emma, Moore points out, grew up in the country, and the area's "extraordinarily rural and quiet" character appealed greatly to them both. Darwin calculated that it was "6 miles from St. Paul's, 8½ miles from [Victoria] station," and "two hours going from London Bridge." Even today, the journey involves a train from Victoria Station to the nearby town of Bromley, a bus from Bromley to Downe (which drops passengers off outside the church where Emma worshipped), and then a walk down Luxted Road to the house. In Darwin's view, that made it far enough from London to deter casual visitors but close enough to be accessible to the friends from London whom Darwin really wanted to see.

It was also close enough to let him stay engaged with the London scientific scene and hear about the latest research in more or less real time. About 14,500 letters sent from Down House survive today, a testament to the effort he spent on cultivating his scientific network. In an age of e-mail and SMS, people might imagine that letters traveled slowly in the nineteenth century, but in the 1840s, Darwin could send a letter to Kew Gardens or the Royal Society in the morning, have it delivered in a few hours, and receive a reply—or a book, seedling, or geological sample—the next day. News traveled fast, but idle gossip and distractions stayed in town.

So Darwin didn't spend his years at Down House in quiet seclusion. He chose it to be close to friends but far from distractions and to have "a place where he can control access to himself," James Moore says. Darwin even altered the property to let him literally "see the world on his own terms." He had a twelve-foot wall built along the north side, and he raised the earth and planted trees elsewhere; he lowered the road that ran outside his house, and, as a final touch, he built another wall from the flints uncovered during the roadwork. "It's all about not having unpredictable disruptions, and seeing who you want to see," Moore explains.

Darwin turned Down House into a kind of scientific field station where he could gather and generate facts. He converted one room into a study and laboratory, added a greenhouse, and took over part of the garden for his research, allowing him to study anything from orchids to barnacles to earthworms. He was a keen observer of the local ecology, and his conversations with pigeon fanciers, dog trainers, and local farmers generated as many insights as his world travels. Indeed, much of the "remarkable body of factual material on which the *Origin of*

Species rests," Darwin biographer Janet Browne says, was drawn from the "commonplace features of Victorian life—letters and small-scale experimental inquiries involving relatively accessible animals and plants."

Darwin said that if he had any special capacity, it was an ability to see interesting things that other people missed and puzzle out their meanings. Down House gave him the space to make those close observations, to pay the attention necessary to observe and ponder things that other scientists overlooked, to think seriously and contemplate. Darwin crafted Down House to amplify his ability to focus. He made it part of his extended mind.

One of the property's simplest but most important features was the walking path. Darwin liked the "narrow lanes and high hedges" around Downe, and many of his letters describing the house and country mentioned walking. "The charm of the place to me is that almost every field is intersected (as alas is ours) by one or more foot-paths—I never saw so many walks in any other country," he told his brother after his first visit to Downe. Soon after moving to Down House, he laid out a walk of his own. Just like garden paths and park trails today, it was formed out of a shallow trench and covered with gravel bonded with sand. The first section was completed in 1843; three years later, he leased a one-and-a-half-acre plot from his neighbor and fellow gentleman scientist Baronet John Lubbock and extended the Sandwalk to roughly a quarter mile. It was Darwin's children who named it the Sandwalk; Darwin himself called it his "thinking path."

For nearly forty years, Darwin's habit was to work "like Clockwork" through the morning, then walk before lunch. Occasionally his children or white terrier Polly joined him; scientists visiting with Darwin accompanied him on the Sandwalk and talked about their work. The area was still very much working land—Down House had a commercial farming operation that helped support the large Darwin family—

and Darwin was notably careful with money, avoiding making improvements to the property unless absolutely necessary. His investment of time and energy on the Sandwalk tells us that Darwin considered it important to have a space to walk and think.

What made it special? The simplest explanation is that, for many thinkers, walking stimulates creativity. The idea that walking helps thinking and can be a form of contemplation has been around since antiquity. The Latin phrase *solvitur ambulando*—"it is solved by walking"—is attributed to ancient philosophers as diverse as Diogenes, Ambrose, Jerome, and Augustine. Buddhists and Christians share a tradition of walking meditation, in which walks along short paths or labyrinths stimulate spiritual reflection and renewal. Walking was an essential tool for eighteenth- and nineteenth-century philosophers. Jean-Jacques Rousseau in Paris, Immanuel Kant in Königsberg, and Søren Kierkegaard in Copenhagen were all famously regular walkers. Kierkegaard declared, "I have walked myself into my best thoughts," and was drawn to walking for both its physical and mental stimulation (benefits that have been documented by modern scientists). So popular was the image of the walking philosopher that Friedrich Nietzsche famously stated toward the end of the nineteenth century, "All truly great thoughts"—including Nietzsche's own—"are conceived by walking." The Sandwalk is one of many examples of paths that philosophers, scientists, and writers have trodden when turning over problems.

Walking stimulates thinking because it offers a break from the hard focused work of writing, composing, or calculating but doesn't completely distract the mind. As Rebecca Solnit puts it, walking "is a state in which the mind, the body, and the world are aligned." As the body moves and the eyes take in novel or familiar sights, part of the mind can still be focused on a tricky issue or stubborn turn of phrase. For someone working on an elaborate problem, a familiar path can occupy some of the mind but doesn't require all of it, providing just enough stimulation to help the subconscious work through dilemmas, test solutions, or break a creative jam.

Throughout his life, Darwin did some of his best thinking and was at his most observant when he was in motion. As a child, he took long walks in the country after the death of his mother. He never recounted what he thought about during these walks—he later claimed he didn't recall—but recent studies have shown that the mental and emotional states of people dealing with grief or recovering from medical crises improve with regular exposure to nature. They don't become happier necessarily, but they do become more resilient and able to face challenges. It's not a stretch to imagine that Darwin learned to find a measure of strength and solace in these walks and that they form the basis of a lifelong connection between walking and contemplation. They also provided the foundation for his discovery as an adult that scientific fieldwork and observation engaged him far more than reading in libraries.

So important was walking to his thought processes that Darwin sometimes described a problem he was working on in terms of the number of turns around his path he would need to solve it. I suspect that formulating problems this way helped him solve them, and that as he trod up and down the Sandwalk, he felt himself walking toward answers.

The Sandwalk could be tranquil and recede in the background when Darwin was deeply focused on a problem. Alternatively, when his mind was stuck or inspiration stubbornly elusive, it offered a physical release and absorbing detail. Underneath the sculptural similarity of a row of trees, there is tremendous variety. In his essay "On Walking," Henry David Thoreau observed, "There is in fact a sort of harmony discoverable between the capabilities of the landscape within a circle of ten miles' radius, or the limits of an afternoon walk, and the three-score-years and ten of human life. It will never become quite familiar to you." There was always more to observe: the progression of the seasons, the cycles of growth and decay, animal migrations, and one's own capacity to shift attention between broad changes in the landscape and novelties within arm's reach. Darwin, with his ability to notice small details and worry them down until they yielded some insight, must have found endless small stimulations on the Sandwalk.

If we apply Darwin's habit of searching for interesting facts in everyday "commonplace features" and look closer at his Sandwalk, we discover that the Sandwalk is a modest but nearly perfect exemplar of contemplative design.

For the last several millennia, a common language, shared by architects, gardeners, and users alike, has helped give shape to contemplative spaces. Builders never cataloged these the way they did architectural styles or garden designs. Then again, nobody seemed to need a contemplative equivalent of the pattern books that architects turned to for help designing facades or houses. People recognized such spaces when they entered them. But recently, landscape architects and psychologists have discovered that whether they're parks or forests, churches or laboratories, medieval Catholic seminaries or modern Zen gardens, sacred groves or academic libraries, contemplative spaces work to quiet the mind and invite reflection by adhering to a few simple rules.

A few years ago, when a client asked Rebecca Krinke to design a contemplative landscape, she did what any professional would do: she looked for works on design principles for contemplative places. There are plenty of studies of specific places, such as Zen gardens, medieval churches, and national war memorials, but surprisingly, she says, "I couldn't really find anything general." Eventually, she found a key to understanding them: the work of psychologist Stephen Kaplan, who's spent decades studying restorative experiences.

Kaplan's abiding interest is in what he calls directed attention, the kind of attention that we need to maintain when we're working on knotty problems or dealing with challenging situations. The ability to focus attention by concentrating on particular things and pushing away distractions—Kaplan calls it inhibition, and it acts to protect directed attention—has always been important, but people need to use it now more than ever. The problem is that evolution selected for humans who pay close, effortless attention to things like threatening or tasty animals, not to traffic, spreadsheets, and business meetings. We don't really have trouble concentrating; the modern world has created a "split between the important and the interesting," Kaplan writes. Combine demands that we pay attention to fundamentally boring things for hours on end with the increasing complexity of technological systems, and you have a prescription for disaster. Many systems collapses, car crashes, and other technological catastrophes begin as a failure of attention or an inability to rapidly redirect attention to new situations.

Studies have documented the restorative value of natural settings and views, and Kaplan identifies four critical features of restorative experiences. First, they're *fascinating:* "They hold your attention without making a demand on your conscious mind," Krinke says. Second, they provide a sense of *being away,* particularly for urban denizens and people who spend most of their lives looking at nature through office or car windows. A third feature is what Kaplan calls *extent:* restorative experiences have to be "rich enough and coherent enough... [to feel like] a whole other world." Finally, they have *compatibility.* They're easy to navigate and make sense of, because they don't throw a lot of different things at us at once.

We all know what these experiences are like. Reading a book can be a restorative experience if the book captures your attention and puts you in another world. Going to the opera or ballet can feel like a trip to another world (people describe themselves as being transported by music or great performances for a reason). These bear a strong resemblance to flow experiences.

Kaplan was interested in understanding why experiences such as reading or walking in the park are restorative. Krinke's insight was that these principles can be applied to understanding built environments and architecture.

Contemplative spaces are purposefully simple. Built spaces tend to use a basic design vocabulary and color palette or employ repetition, while gardens and parks use a few well-placed plants or trees to catch the eye. Muted sounds and shade and shadow dampen the visual and auditory impact, encouraging visitors to relax and focus. *Simple* doesn't necessarily mean "stark" or "blank"—even the hermit's cell will have a window and a scroll or a cross. More often, contemplative simplicity is like that of a museum: the space is stripped down to allow a few objects to come into focus.

This simplicity may be a strategy to make an environment more tranquil, or it can be used to frame attention more intensely on a specific place, object, or ritual. Our awareness of a grand theater's decor recedes when the house lights go down and the play begins; a monumental gallery will use clean blank space or pinpoint lighting to funnel visitors' attention to a painting or sculpture. Cathedrals become warm and intimate when lit with candles during evening services; their vast ceilings fade into shadow, allowing the eye to see worshippers in a smaller, more intimate space.

Contrasts are another feature of contemplative spaces. The small cloistered garden shadowed by a mountain, the stone path that leads to the water's edge, the narrow dark passage that opens onto a sunlit square, the temple that meets the sky—all these bring together microcosm and macrocosm, contrasting elements such as confining dark and expansive light or human and nature. Arranging spaces into contrasts of natural and built, moving the visitor from the small to the large, or from the dark to the light, turns transits into little pilgrimages.

The best contemplative spaces "often have to offer a sense of being away" with "a sense that they're connected to larger systems or parts of your life," Krinke says. Visiting a cave that was sacred thousands of years ago to a nameless, now-forgotten people may offer lots of mystery but not much sense of connection. Urban parks, by contrast, are restorative precisely because they're physically close to visitors' regular lives but still offer something very different.

Indeed, creators of contemplative spaces have long intuited that people find untamed, absolutely wild spaces threatening, not restorative, and are reassured by traces of human presence on the landscape. The mountain with a trail or the desert with an oasis and a traveler's hut in the distance feels different than the open ocean or a dense jungle. Paths or built objects help make the space comprehensible; the path makes a narrative, the gazebo or shrine provides a destination, and the act of movement becomes a pilgrimage. It is easy to overlook the critical role that a sense of location—a certainty that you know where you are and where you're headed—plays in the feel of a place. We celebrate the freedom to wander, but not having a destination is most definitely not the same thing as being lost—the ability to drift aimlessly implies that you know where you are. (Imagine walking through a city you know well, spending an afternoon with no fixed destination but with a constant awareness of your location. Now imagine being lost in an unknown city. Your path may look exactly the same on a map, but it will feel very different in the moment.)

The Sandwalk contained all the features of a contemplative space. It was purposely simple, just a broad oval running on the acreage rented from his neighbor John Lubbock, one side opening onto a meadow and the other plunging into woods. Darwin didn't add architectural follies or other diversions. He kept it simple, clearing here, planting trees and shrubs there, but never trying to compete with the great old elm tree in the hedge (he noticed it on his first visit to the property) and keeping the landscaping in tune with the countryside.

Within that simple layout, though, Darwin designed in broad contrasts. His habit was to walk out along the southern edge of the property, where the hedges and fences were low, and one could look out over them to open fields and more forest and hills beyond. This section ended at a small gazebo that the family called the Summer-House. He would then turn around and head north along a second, narrower and darker path, shaded by trees, eventually coming back out in the light.

It was also a pilgrimage. To get to the walk, Darwin would go out the back door of the house and follow a straight and broad path past greenhouses and gardens to a wooden door in a high hedge. The door opened onto a meadow and the top of the Sandwalk. Though it was a mere thousand steps from the house, it "seemed to be a very long way from the house," Darwin's granddaughter recalled, as the hedge "quite cut you off from human society." Darwin described Down House as "absolutely at the extreme verge of the world." He placed the Sandwalk on the edge of that extreme verge and made it a retreat from the domestic bustle of the respectable country house.

The Sandwalk's beauty is natural but not wild. Darwin planted dogwood, hornbeam, and half a dozen other species of trees along the walk on his property, and he landscaped the Lubbock rental, installing trees and shrubs on one side and a fence on the other. "It was Darwin's greatest piece of horticultural engineering," Jim Moore tells me. Beyond that, though, the simplicity of the walk, the combinations and contrasts of darker enclosed and brightly lit spaces, the shifts in perspective from the close at hand to the very distant, the blend of human and natural—it brings together all the classic elements of restorative environments and contemplative spaces.

Is it unreasonable to describe the Sandwalk as a form of information technology, a tool Darwin used to focus his mind? He outsourced the task of keeping track of his progress to a pile of stones, which he would move one by one as he passed. (His children sometimes hid the stones to see how deep in thought he was.) He completed eighteen books and monographs at Down House, including *The Origin of Species* (1859), *The Descent of Man* (1871), and *The Expression of the Emotions in Man and Animals* (1872). He spent thirty-six years of his life with the Sandwalk. If he went out on it three hundred days per year and walked an average of two miles each day, that's over twenty thousand miles on his thinking path; enough for him to circumnavigate the world a second time and change the way we see the world.

As Moore talks about Down House, I'm struck by how patient and certain Darwin must have been when he laid out the Sandwalk. I live in a working world in which projects last for only weeks or months; deadlines are tight, the market is competitive and unforgiving, and if you don't get there fast, someone else will. The world Darwin lived in seems completely alien, while the natural world he described, the life of endless competition and struggle, is instantly familiar. My hypersmart friends work immensely hard, and virtually none of what they create lasts—they devote their lives to building things their competitors will promptly tear down and replace. Even the best ideas and most valuable expert knowledge has a vanishingly short shelf life: five years from now, the most bleeding-edge technical expertise and breathtakingly valuable patents won't be catnip for venture capitalists but feedstock for some bargain-basement commodity technology churned out by a factory outside Shanghai. Likewise, our personal lives are marked by transience. You might be in the same place in three years, or you might be in Seoul or Dubai or Boulder, chasing the next opportunity. Even if you don't move, you feel as if half your existence is in the Cloud. And if you do leave, by the time you return home, the place will have changed beyond recognition.

Darwin, by contrast, planted those trees believing that he would watch them for decades and that their growth would pace his elaboration of his theory of evolution. When he rented the land from Lubbock, he looked forward to having "one sheltered walk" and the "amusement in tending and pruning the trees." The "tangled bank" that he describes at the end of *The Origin of Species* grew in miniature along the path he marked and walked. Down House was a world that alternately stimulated, shielded, and supported him for decades.

It suddenly seems like a very different reality. How can we possibly hope to reproduce anything like that for ourselves?

The answer is, most of us can't. But we can use the design principles behind contemplative environments to create restorative experiences and interactions even in small, temporary, and unexpected places. And we can learn to use tools to create fascination, removal, extent, and compatibility.

Krinke's and Kaplan's work explains why I've learned to treat airplanes as contemplative spaces. I focus in those cramped quarters on those elements that make it a restorative environment. I push the plastic food, the stressed fellow passengers, the little dramas over possession of the overhead compartment out of my mind. I let the flight become my own Sandwalk.

The foundation for that conversion is in my own deep associations between flying, adventure, and discovery. During my childhood, my family migrated between the States and Brazil as my dad worked on his dissertation. My memories of everyday childhood life in Brazil are fuzzy, but my memories of *traveling* are vivid. The nights we spent on buses to Mato Grosso and Ouro Prêto, rumbling from the tropical coast to the interior plains; the flights from Rio to Buenos Aires or Bogotá; the view of the Amazon and the Andes under the wings of our plane—all are easier to recall than the apartments we lived in or the parks where I played. For me, flying retains a kind of glamour and excitement—to use Kaplan's term, a *fascination*—that the realities of delayed planes and overpriced food can never completely dull.

I fell in love with flying in the late 1960s, back when the idea of affordable international travel was still a novelty and the world was not yet overrun by global brands and chains. The first transatlantic passenger jets started operating ten years before I got on a plane, and nonstop travel from North America to South America was still a rarity. There's a parallel to the culture of walking in Darwin's time. When the young Darwin was going on his solitary walks after his mother's death, the Romantic idea of walking as a form of self-discovery was still fresh, and Gilbert White's *Natural History*—which helped spark public enthusiasm for botanizing and collecting—was about thirty years old. Cultural associations nudge me to seeing both planes and paths as contemplative spaces.

The physical environment of the plane, I now observe, has all the features of a contemplative space. My narrow personal space is crowded but orderly, and everything I need is within reach; it's the road warrior's version of a Zen dry garden, comprehensible and comprehensive, physically small but infinite in intellectual extent.

Much of the time, my attention stays within this little world. On night flights—my favorite kind by far—the cabin is dark, illuminated only by pools of light around my fellow workers and insomniacs and the occasional flicker from movies. It's no forest, but the mix of light and dark, the blend of quiet and occasional stimulation, helps my mind settle and be more focused.

Even the sonic environment helps me concentrate. I work with music, but I'm still slightly aware of the plane's engines; I can feel them as much as hear them. Audio engineers will boost low-frequency sounds, which people associate with large spaces, to make a room feel bigger than it really is. (Moviemakers also add low rumbles to the sound track when a scene moves from small to large spaces.) I suspect that while my physical space is clearly narrow and constrained, at some level, the deep rumble of the plane, the complex symphony caused by the shock of subzero-temperature air against the aluminum fuselage, makes me feel I'm in a more expansive space.

And being offline makes a difference. It's the digital expression of my being removed from things, one more little way that the transit becomes a pilgrimage. Knowing that I can't get to my e-mail or the Web, I don't even try. I'm relieved of whatever twinge of desire I might have to check the news. The delicious junk food of digital distraction is out of reach. Even more delicious is the feeling that other people can't reach *me*.

Of course, I'll be back online in no time. For now, though, it seems a long way off. Time suspends and stretches on a long flight. We'll land eventually, but during those compressed hours of nighttime and slow dawn, I have the sense of being at once aware of my deadlines and unconstrained by normal duties. I'm very much in my working life but apart from everyday life.

Just as important, after days of frantic rushing, my body *has* to stop, even as my mind continues to race and I keep moving across time zones. My psyche still hums from the day's preparation and packing, despite the fact that I have the chance (and the enforced need) to just be *still* for ten hours, to feel time unwinding slowly. This alternation between fast and slow, between movement and stillness, between physical immobility and mental racing creates a state where action and contemplation don't alternate like day and night but join together.

We think of breaks or rests as the equivalent of an Off switch: either we're working or we're not. For driven alpha personalities, the need for rest is a lamentable bug in the human operating system. But restorative experiences don't require your mind to switch off. They aren't disruptions for the working and creative mind. They create another, quieter but equally valuable state in which the creative mind is still working, just in a different, less directed way.

Sinking into a more reflective and thoughtful mood takes time. It's not something you can do instantly. When I meditate, it takes a few minutes for my body to calm down and be still; only then can I begin work on clearing my mind. Learning how to create contemplative states and restorative spaces also takes time. I'd been flying for years before I started working creatively on planes. But it's worth taking the time to immerse yourself in restorative environments for a few minutes, to learn to convert pauses and breaks into opportunities to let your mind recharge, to shift into a lower gear rather than turn off entirely.

It's easiest to hear these changes of key over the course of a working day, but you can see them play out over the course of years too. Darwin spent years on the Sandwalk; he observed the difference in the light between morning and afternoon, noted the changes of the seasons, and, over forty years, watched the trees he planted grow. His intellectual trajectory and his combination of travel and contemplation demonstrate how, over decades, a creative life mixes active and reflective times, travel and rest, the novel and the familiar, an ancient desire for solitude and a very modern sociability. Darwin circulated through multiple spaces during his days, and he was careful about what places he occupied and what he let occupy him. The results speak for themselves.

We have an impoverished vocabulary to describe restoration. The term *distraction* is sometimes pressed into service as a synonym for it, but there's a big difference between watching a YouTube video of dogs playing poker while IM-ing with several friends and dealing with an urgent call from the office, and going on a hike. Restorative activities and environments occupy your conscious mind, leaving your unconscious free to work without deliberate effort and with an understanding that the pressure isn't on as much.

And if short breaks aren't enough, you can take every device you normally carry, everything with wi-fi or a screen, every single source of distraction, and turn them all off for a day. Really. You can.

SEVEN

REST

One evening, disentangle yourself from wires and wirelessness. Switch off the fire hose of alerts and updates. Insulate yourself from the tug of the thousand tiny peripheral connections that poke at you during your normal day. Put an away message on Facebook. Switch off your wi-fi network. Set your cell phone to vibrate, and don't put it back in your pocket; instead, leave it on a table. Leave the tablets and games to charge. Put your laptop in your bag. Then put the bag in the closet.

For the next twenty-four hours, don't go online, don't check your e-mail, don't use anything that has a screen. Find that book you started last month (or last year) and finish it. Catch up on magazines. Find out what friends have actually been doing, not posting. Cook a big meal with them. Find the candles and corkscrew, dust them off, and use them. Work on the car, or take apart and clean your bike. Whatever you do, let it be something that's interesting, engaging, and makes you feel more firmly part of the real world.

At first, it'll be hard. If you're like me, you're accustomed to the Internet always being available, and being without it will seem pointless and counterproductive, even dangerous. What if someone's lying in a ditch and tweets that he needs help? How will you feel if you miss it? (Don't laugh. In emergencies, people have been known to tweet first, *then* call the police.) What if something happens somewhere in the world, and you don't hear about it? Rationally, you know this kind of anxiety borders on the absurd. But the fact that you feel that way is a sign that you really need this break.

The next evening, if you rush back to your computer and feel like the glow of monitors is as welcoming as a cozy house on Christmas Eve, don't worry. You're just like everyone else.

But next week, disconnect again. It may still be a challenge, but it will be a little easier.

After two or three times, you'll start to notice some changes. Unless you're a reporter, currency speculator, or emergency-department doctor, you'll discover that your world doesn't fall into chaos when you go offline. An amazing amount of e-mail is either fluff or stuff that can wait; we just assume that all messages are urgent. People who really need to reach you still can; in today's world, an airplane is about the only place a person stands a chance of being *really* unreachable.

You might feel your mind slowing down a bit, but in a good way. Some of the cognitive sediment stirred up by juggling work, personal life, and virtual distractions is starting to settle. And the stillness that's left, which people usually assume is a terrifying boredom that has to be filled with something, actually isn't bad after all. It's the feeling of your extended mind tuning up, your attention rebuilding, the balance between the human and high-tech parts of you righting itself.

Welcome to the digital Sabbath.

Like many technology innovations, the digital Sabbath movement started in Silicon Valley. The term was first used in a course on "the art of aligning your inner and outer lives and living life according to your values" taught by Anne Dilenschneider, a psychologist and Methodist pastor who works with nonprofits and clergy, and Andrea

Bauer, an executive coach who works with Silicon Valley CEOs and managers.

Each had seen her share of people working ten-hour days, living in a constant stream of e-mail and meetings, losing the ability to step back and reflect. Even clergy were treating churches like startups, and they faced pressure to fund-raise, develop new programs, put their sermons in PowerPoint, and grow their congregations. "We wanted to do a class that helped people reconnect with themselves," Dilenschneider recalls, speaking to me from Fargo, North Dakota, where she's completing a clinical psychology residency and working as a pastor. Inspired by Julia Cameron's idea of an "artist's date," she and Bauer told students to spend a day unplugged, to take a break from the world of work and the endless tug of e-mail, turn off their pagers and PalmPilots (cutting-edge technology when they taught the class in 2001), and spend the day doing consciously low-tech things.

The first Sabbaths were "challenging," Dilenschneider recalls, but they "led to some great conversations about why can't you unplug, whether you're really so indispensable that you can't unplug, and that the world is not going to end if we're not online." Digital Sabbaths and similar events—they go by names like Screen-Free Week, Offlining, and Disconnect Revolution—have grown more popular since Dilenschneider and Bauer first offered their class, and they've attracted an interesting mix of wired but thoughtful people.

I interviewed a number of digital Sabbatarians to learn why they started unplugging, how they do it, and what benefits it brings. They're writers, consultants, lawyers, entrepreneurs, graphic designers, engineers, educators, even advertising executives. Their work involves juggling multiple projects and clients and requires creativity, self-management, and self-motivation—in a word, they have to do a lot of multitasking, and they have to be self-focusing. They're digitally savvy, but they often have strong analog interests as well. One of the earliest proponents of regular breaks from connectivity is University of Washington professor David Levy; he has a PhD in artificial intelligence (from Stanford) *and* advanced training in calligraphy and bookmaking (from the Roehampton Institute in London). Others are beer makers or cooks or extreme-sport athletes. They're eager to share their stories but don't want to be seen as antitechnology or antimodern; almost all of them said something like "I'm not a Luddite, but—" when talking about their digital Sabbaths.

Some people discover digital Sabbaths by accident. For example, Martha Rock uncovered the benefits of being completely offline when her family signed up with a bad Internet service provider. She had recently left her job as in-house counsel with a Silicon Valley tech company and become a preschool teacher, but she still found herself "overwhelmed by the volume of electronic stuff." Her family was disquietingly hyperconnected. She'd see her teenage son studying with "a book in his hand, YouTube videos going, listening to music, and texting" and her husband checking his BlackBerry when they went on dates. Dealing with the ISP was a trying experience, she recalls: "They made me cry," she says with mock sadness. She hadn't meant to start taking digital Sabbaths, but a faulty Internet connection gave her a welcome break from parents hunting for volunteers to run school events and a school principal who expected her to be accessible online.

David Wuertele stumbled into it thanks to his toddler son. David is an engineer with electric-car company Tesla Motors—its high-performance electric roadster is a must-have accessory for green venture capitalists and socially conscious CEOs—and he has considered being online to be "like breathing" since his student days at Berkeley. He started going offline when he spent Saturdays with his year-old son. At first, he would bring along a tablet computer to the park, but he noticed his son would want to do something, and "I was telling him to wait while I finished an e-mail or a paragraph." He worried that his son would feel neglected and that his e-mail would distract him from those small moments that make parenthood special. So he started leaving the computer at home, switching off his phone, and bringing a book to read when his son napped.

Shay Colson first went offline when he took his wife to Bali for a monthlong honeymoon. Shay's the sort of techie who can tell from your phone's notification sounds if you're an Android, Windows, or iPhone user, but after years studying information sciences at Syracuse University, he was ready for "a chance to reset" his communication with others and refresh his relationship with technology. When they started their monthlong trip at Seattle-Tacoma airport, Shay and his wife carried real guidebooks, paper tickets, and printed reservation information. Their only electronics were a digital camera and his-and-her Kindles. Since he had no way to post a picture of himself on Facebook along with the words *OMG I'm snorkeling and tweeting at the same time LOL,* he found it easier to stay in the moment, "to be with my wife, be fully aware of what we were experiencing."

Tammy Strobel's digital Sabbath is part of a larger effort to live more simply and meaningfully. A writer and Web designer, Tammy had been experimenting with voluntary simplicity for the last couple of years, shedding her car and cell phone and moving with her husband into a custom-designed 150-square-foot house in Oregon. Tammy started her digital Sabbaths when she felt herself using e-mail and the Web to "avoid hard work, or the fear that 'OMG my writing sucks, I'm a failure.'" When she felt them acting less as tools and more as self-distractions, she started "powering down all my gadgetry" on weekends. E-mail, Internet, wi-fi would all go off, to be replaced by books and time with her husband.

Christine Rosen, an author and senior editor at the *New Atlantis,* started taking digital Sabbaths after writing about how technologies like DNA fingerprinting and GPS subtly shape how we experience time, work, and family life. Her writing made her especially aware of the ways the technologies can be both necessities and disruptions. One day she noticed that when her husband brought his laptop into the living room, "the kids were just *drawn* to it." So computers were exiled from family spaces. She also noticed her time online was changing the way she read. "I was always a big reader," she says, but she realized, "reading [online] in a scattered fashion wasn't giving me much pleasure." She installed Fred Stutzman's Internet blocker Freedom on her computer and started setting aside time in the evening to immerse herself in books. From there, it was a short step to a digital Sabbath.

In their original challenge, Dilenschneider and Bauer suggested planning Sabbath activities in advance, clearing away clutter and household tasks so they wouldn't be a distraction, and inventing a short ritual to mark the end of the day and the return to normal life. A decade later, in the wake of social media, iPhones, Xboxes, and Cloud computing, what do people who take digital Sabbaths actually do? What do they turn off, how do they spend their days, and what do they get from being offline?

The timing is often easiest to define: sundown to sundown, or an entire weekend day, are the most popular daylong choices. It's easier to start and maintain a digital Sabbath if it has the predictability of ritual.

Just *what* devices you unplug requires a little more thought. Nobody I spoke to unplugs absolutely everything: "I could never go the full Amish," as one friend charmingly (although inaccurately) put it. Some people follow technical standards; they avoid everything that has a screen, or put away everything with a wi-fi connection, or turn off anything that has a power button. These rules are easy to remember and enforce (especially valuable if you have children) and can clarify the boundaries between normal life and digital Sabbath. In other cases, you might want to get away from especially distracting services and devices; some people shut off cell phones and computers and make the kids hand over their handheld games but still allow an all-family Guitar Hero smack-down on the Xbox.

Many digital Sabbatarians choose what to unplug based on the psychological effects of technologies, not their technical properties. Writer and consultant Gwen Bell, for example, puts away "anything that in my experience has a quality of addiction to it." Shay Colson thinks of being online as the "Pavlovian experience" of "being available to inputs that you don't directly control." By this definition, watching broadcast TV counts, while cueing up Netflix or diving into his DVR does not. Carrying a cell phone with the ringer and e-mail notifications switched on is being online; putting the phone in quiet mode is offline. "It's about control, mental energies, and your own inputs and outputs," he concludes. Making these choices requires looking at your normal technology use, thinking about what devices are most psychologically demanding and distracting, and deciding what interactions have the potential to become addictive. The first step to successful disconnection is introspection.

Unfortunately, the next step is misdirection. We're accustomed to thinking that going offline means missing out and that social life today is a flat network rather than a hierarchy, so a person who purposely disentangles himself from the online world can come across as hostile and antisocial. Martha Rock was shocked at the response she got when she told people she was spending less time online. "People would ask, 'Can you run this bake sale?' and for my own health and well-being, I'd explain why I couldn't. I thought I was being so authentic, sharing my vulnerability," she says. "Boy, was *I* naive. People. Freaked. Out. Oh, they *so* didn't want to hear it. It was viewed as *hostile.* Or *precious. Oh, look at* us, *trying to have mental health.* Or they'd just be like, *Fuck you. Who's going to do my bake sale?* It really rocked my world." "You don't need to tell people you're going offline, because that gives them a chance to object," Shay Colson says. Now, he admits, "if people say anything, I say, 'I didn't have service.' "

You don't have to make yourself unavailable to people who *really* depend on you. As members of what sociologists call the Sandwich Generation—those responsible for both children and aging parents (in earlier years, social scientists called such people "women")—lots of adults today are uncomfortable with the idea of being completely out of touch. The aim of a digital Sabbath is not for you to be irresponsibly inaccessible but for you to filter out unnecessary distractions, like the manufactured urgency of a client's latest brain wave or a surge in tweets, while still being accessible to field the call from the nursing home or elementary school. Distracting yourself with anxiety over being unreachable is no better than distracting yourself with the latest LOLcat picture.

Think about what kinds of activities will engage you. Reading books is the most popular act of protest against the incursions of digital distraction. Writing in nice journals, reconnecting with the physical pleasure of committing ink to a page, treating the slower pace and greater permanence of ink as an invitation to deliberate over words is a close second. Cooking requires focus, rewards creativity, is intensely tactile, and can be solitary or social. Sewing and knitting have the same benefits. David Wuertele, who describes himself as an "engineer's engineer," takes himself offline and works on building an industrial-grade home brewery, complete with stainless-steel custom-machined parts he's designing himself. (There's a religious precedent for this: the monastic Thomas Merton once declared, "I love beer, and by that very fact, the world.")

All these activities are complicated but doable, mentally absorbing, engaging to the senses, and immediately rewarding. Almost all can be pursued either alone or with friends, depending on your mood and circumstance. They offer some of the stimulation of being online and engaged with the digital but none of the distraction, none of the sense of having to divide yourself and your attention between two worlds. They're restorative, flow experiences.

Be patient. It's normal to struggle at first with an extended period offline, and it's only over time that your mind will learn to slow down and take advantage of its unaccustomed freedom. Lots of people report being nervous the first time they turned off the Internet and e-mail for a day; some felt like they were going into withdrawal, and others had trouble shaking their online habits. To satisfy her craving to post updates, Gwen Bell "paper tweeted" (otherwise known as "wrote things down") during her first sabbatical. Everyone worries about having an urgent e-mail sit too long in the in-box, missing a call, or not hearing some important news story. But after several collective years of Sabbaths, my interviewees say that they've *never* overlooked an important communication, and that anyone who really needs to reach you *will* find you.

Don't think of it as a daylong vacation. As Dilenschneider reminded me, the Jewish Sabbath gives believers a chance to "remember you're made in the image of God, that your value comes not from what you *do* but what you *are.*" But this requires making the effort to discover what you are. It "creates an obligation to know yourself, to wake up and be aware." This means stepping out of the rushing streams of commerce and connectivity and into a very different kind of space, one that replaces speed with stillness, reaction with reflection. It's an invitation to treat the digital Sabbath as an opportunity for spiritual renewal.

Maybe this idea feels alien to you. If so, you're hardly alone; the digital Sabbath movement itself is a little ambivalent about its religious roots. Dilenschneider and Bauer held their 2001 class in a Lutheran church. The Sabbath Manifesto, a set of ten principles to help "slow down lives in an increasingly hectic world," published in 2009, was produced by the Reform Jewish group Reboot. Yet although its suggestions to turn off devices and connect with loved ones, light candles, drink wine, and eat bread echo Jewish tradition, the manifesto eschews religiosity in favor of inclusivity.

Most digital Sabbatarians I interviewed are not religious; some described themselves as "mildly observant" or "spiritual" rather than religious. It echoes the "I'm not a Luddite, but—" caveat that I hear so often; people are drawing boundaries around their behavior. Intrigued, I asked Morley Feinstein, chief rabbi at the University Synagogue in Los Angeles, what this means. Feinstein has led congregations in the Midwest and California, and he looks so much like the archetypical rabbi that he could play one on TV. (In fact, he once had a cameo on the sitcom *Curb Your Enthusiasm*.)

Feinstein explains that many people "end up doing traditional things, but God forbid they call themselves religious." It's about self-perception and about how you think others will perceive you, as Daryl Bem puts it. Religious is what your grandparents were. Being "spiritual is okay," Feinstein continues, "but call yourself religious" and you assume others "immediately think you're from Borough Park," a Hasidic neighborhood in Brooklyn.

Feinstein sounds a bit exasperated, yet he is sympathetic to people who need a break from their devices but don't possess the language to explain that to themselves. "When people say, 'I'm turning off my iPhone for the Sabbath,' they're not saying, 'I need to reclaim my neurons,'" he points out. They "gain spiritual, emotional, health benefits from it. It's a traditional observance. That's hard to admit if you're ultra-hip." And this movement is part of a longer history of reimagining the Sabbath. Other events in the religious calendar might be dismissed as artifacts of an agrarian age or as too demanding for the modern age, but Shabbat is "the only holiday that's written in the Ten Commandments, so it's never lost its meaning and power." It's unavoidable. So, he continues, "every fifteen years, there's a push to make it more meaningful, more accessible." The digital Sabbath is just the latest effort to update it.

I'm sympathetic to the idea that the digital Sabbath should welcome everyone, and it makes sense that it should not be too concerned with doctrine or get bogged down with the minutiae of observance. An essential part of contemplative computing is learning what works for you.

Older writing about traditional Sabbath has some valuable ideas, regardless of your faith. But it may not be obvious that religious texts can help you think about technology, and maybe you find the idea of diving into Jewish writing about the Sabbath a little disquieting.

I get you. I'm entirely nonreligious myself, and most of my friends are either cosmopolitan skeptics or too busy on Sundays (or Saturdays) with work, kids' sports, and errands to go to church (or synagogue). My approach to religion is analytical and anthropological: I observe what believers do and appreciate how it enriches their lives, but I can't experience it myself. So you can imagine my shock when I picked up Rabbi Abraham Heschel's 1951 book *The Sabbath: Its Meaning for Modern Man* and found it speaking to me as profoundly as anything I'd ever read.

The book is not a guide to observing the Sabbath but an analysis of its deeper meaning. Sixty years after its publication, *The Sabbath* remains a jewel—small, multifaceted, and brilliant. It sealed Heschel's reputation as one of the twentieth century's most important Jewish theologians. Daniel Nevins, the dean of the Jewish Theological Seminary in New York, explains that Heschel is still "revered for his poetic writing." "He's absolutely treasured," Morley Feinstein agrees, then adds the ultimate compliment: Heschel is "the Einstein of modern theology."

The comparison turns out to work on two levels: Heschel's writing displays a singular genius, and, like Einstein's theory of relativity, *The Sabbath* is about the nature of time and space and how we relate to them.

For all its dull familiarity, Heschel argued, the Sabbath is organized around some radical ideas. It's profoundly egalitarian: everyone, including servants and the poor, has the right to enjoy the day. (Even beasts of burden get the day off.) It's innovative in assigning sacredness to a *time* rather than a space. Ancient people worshipped local gods who lived in sacred groves or forests or mountains. The book of Genesis, by contrast, describes the world as good, but only the Sabbath day is considered holy. Judaism, Heschel concluded, "is a religion of time aiming at the sanctification of time," and the Bible encourages readers to recognize that "every hour is... exclusive and endlessly precious."

Foremost in Judaism's "architecture of time," the pinnacle of its rituals and commemorations, is the Sabbath. "The meaning of the Sabbath is to celebrate time," Heschel said, and even among Jewish rituals, it is uniquely focused on giving practitioners a taste of the holy and eternal. The timing of most events in the Jewish calendar is set by nature or history, but the Sabbath's rhythm isn't tied to the lunar cycle or seasons; it mimics the cycle of Creation itself. It encourages believers to see time and space the way God made them. Indeed, Heschel argued, "the essence of the Sabbath is completely detached from the world of space." The Sabbath offers an opportunity to "become attuned to holiness in time" by building a "palace in time... made of soul, of joy and rectitude... a reminder of adjacency to eternity."

For Heschel, a descendant of a highly respected family of Eastern European rabbis and a Berlin-trained philosopher who fled Nazi Germany on the eve of World War II, the vision of a religion that lived "in the realm of time" and could survive the loss of mere places must have been powerful and comforting. But Heschel's ideas about time, renewal, and the relationship of the Sabbath to normal life can deepen our understanding of how to make the most of digital Sabbaths.

Sabbath time, with its "adjacency to eternity" and disregard for the pace of politics or commerce, could not be more different from the time when news and information flows, financial transactions are conducted, and other online activities occur—that is, real time. The term *real time* first appeared in the late 1950s, coined by computer scientists designing systems that could analyze and respond to information as quickly as they took it in. It started as the effort to create computers that could *mirror* reality, but it has since acquired the power to *change* reality. There are now huge, fast fortunes to be made in real time, in executing financial trades a few microseconds faster than the competition, in bringing products to market more quickly, in giving social-media users the seductive feeling that they can know what other people are thinking or doing *this instant.* In this world, if you go offline, even for a little while, you're losing money and losing out.

The most consistent thing about real time is its inherent instability. As computer systems get faster, real time speeds up; as the systems are woven ever more tightly into the world, the demands of real time intrude on ever-larger parts of our daily lives. It's imperative and insistent. It always wants to go—wants *you* to go—just a little bit faster. It's unlike the standardized time of the nineteenth-century factory and railroad, which treasured predictability and was calibrated by the precise clocks of astronomical events. It's very different from the rhythms of nature or our biological clocks. And it could hardly be more different than the long, majestic, eternal time scales of creation and eternity.

Trying to keep up with real time can exact a high price. It creates services to make online life and communication quicker and more frictionless—one-click purchasing; texts rather than e-mails. It disrupts life to make life more seamless. Trying to live at the speed of finance and commerce and communication forces us to focus on the present moment—the present instant—and erodes our ability to slow down and think. "Minds, organizations, cities, entire societies all need time to integrate and process new ideas," futurist Anthony Townsend says. "If you think you have to constantly, instantly react, rest and contemplation and deliberation—the ability to think about what you're doing—disappears." Relentless, constant exposure to real time, he argues, "destroys both decision-making and contemplative ability."

So Heschel's vision of the Sabbath as an invitation to step into a "palace of time" that hints of eternity and experience time set to the clock of creation rather than the speed of light is more valuable today than ever. In ways that Heschel did not predict but would have appreciated, today's digital devices and virtual spaces create a terribly intimate "tyranny of things." We labor, as people always have, "for the sake of things." But, Heschel noted, "possessions become the symbols of our repressions, jubilees of frustrations.... Things, when magnified, are forgeries of happiness, they are a threat to our very lives." We unconsciously treat computers like people, and our mobile devices act like toddlers: they're super-responsive, simultaneously eager to please and oppressively demanding, always on, and insistent that we pay attention to them. Today, we live our lives with and through objects that compress our experience of time, that replace the rhythms of days and bodily metabolisms with the 24-7 real time of networks and markets. Heschel's warning that "we are more harassed than supported by the Frankensteins of spatial things" seems truer than ever, especially now that those Frankensteins have begun to demand their creators' attention and love.

What *The Sabbath* offers is a day in which it's acceptable to step away from all of that. For one day a week, it's all right to "collect rather than to dissipate time," to "mend our scattered lives." This is where digital Sabbatarians are headed. They intuit that behind the language of disconnection lies an opportunity to do something very profound: to mend one's relationship with time. It's a chance to learn how to collect rather than dissipate time; an invitation to experience a more majestic, mystical time that lengthens one's capacity for attention and for presence; and an opportunity to make meaning in life.

Some regard the digital Sabbath as impractical or ineffective, the equivalent of a crash diet. Starvation, critics point out, is not a basis for healthier patterns of consumption. You can't change your digital consumption habits by switching everything off for a few days, any more than you can reset your body's metabolism by not eating for a day. To these commentators, the digital Sabbath is the sort of extreme measure pushed by miracle-diet fads and weight-loss gurus, the electronic equivalent of a colonic cleanse.

If you think information is like food, then this is a perfectly reasonable critique of digital Sabbaths. But that's the wrong way to think about it. Periods of holy fasting, like Ramadan in Islam, Ash Wednesday in Catholicism, and Yom Kippur in Judaism, aren't diet plans; the intention is completely different. The ritual abstention from food is done to build piety, redirect one's attention away from worldly concerns, increase self-discipline and restraint, and cleanse the body and purify the soul, not to shed sinful pounds.

Likewise, the point of observing a digital Sabbath is not just to reduce your virtual weight index, as Daniel Sieberg calls our accumulation of digital devices and online identities. Digital Sabbaths offer you a chance to exchange the challenges and stimulation of the Web for those of books and landscapes; the gratification of likes and follows for the pleasures of cooking and craft; the rewards of friending people a world away for those of connecting with people nearby.

In interviews, digital Sabbatarians explain how their breaks improve their everyday lives and refresh their online relationships. Discovering how few important e-mails they miss while offline is a revelation. E-mail has a way of making itself seem urgent. But when you step away from it, you can discover that the pressure is almost completely artificial. "I have never missed out on anything by taking a day off," Christine Rosen says. Shay Colson marveled over how little substantive mail had come in during the month he was on his honeymoon. "There were only a few messages worth reading," he recalls. "The signal-to-noise ratio is so out of proportion, it's almost overwhelming."

The Sabbath pares away peripheral connections. People don't cut themselves off completely, but many talk about communicating selectively. Martha Rock says her communication "became more personal by default" when she started taking Sabbaths. "At first, people were pissed," she continues. "They were like, *What else? Do you have a horse-drawn buggy?* But the ones who really matter know how to get to you." Tammy Strobel found that scaling back her e-mail time let her "write things that are more meaningful." E-mail is useful, "but not if I'm checking it every fifteen minutes, and not if I feel really distracted and rushed."

The Sabbath increases your ability to concentrate on cool intricate tasks, to experience and appreciate the uniqueness of particular moments, to focus more on the people around you. Paying attention is critical for relationships. "Developing a relationship is all about being in the moment, being mindful about how you conduct yourself," Tammy Strobel argues, and that sense of presence is hard to maintain when you're wondering what's in your in-box. Because you are freer to spend time absorbed in things you love to do, activities that are engaging and fun, doing things offline is restorative, not exhausting.

Reducing the spam and peripheral connections creates unbroken and undistracted blocks of time. As Gwen Bell puts it, most of us are accustomed to "subdividing" time "into smaller and smaller increments," and dividing it yet further between people on the phone and tasks on the computer. We expect this to make us more productive, but it has the opposite effect. Spending long stretches of time on one thing illustrates how inefficient switch-tasking actually is. It's "astounding how much time is in the day when you don't spend it in thirty-second chunks," Shay Colson says. "Days are long, and you can do an incredible amount with each one. We all know this, but it's easy to forget, especially when you're distracted with whatever you're doing online."

Enforcing distinctions between online and offline time makes it easier to finish tasks and to maintain a better separation between work and regular life. "I feel more free" with the clearer boundaries that the Sabbath sets, Tammy Strobel says. "I don't have to be tethered to my e-mail." Rather than stay online until bedtime, she'll "power off and focus on reading or talking with my husband." Christine Rosen finds that during a digital Sabbath, "I give a lot more attention to how time passes in a day. I have more awareness of what I'm doing. It's not a sacrifice. It's a pause before a busy week begins."

Abraham Heschel saw the Sabbath as a counterbalance against modern "technical civilization," a way "to work with things of space but to be in love with eternity." In Heschel's reading of Genesis, the seventh day was when God created happiness and tranquillity and perfected the universe; it was not the end of Creation but its culmination. Heschel argued that we're obliged to re-create that happiness and tranquillity. "Rest without spirit," he warned, is "the source of depravity."

The Sabbath isn't a day for mindless leisure or diversion. For Heschel, Sabbath rest was not *passive;* it was *active.* In *The Sabbath,* Heschel was impressively silent about whether any particular activity should or shouldn't be undertaken on the Sabbath. You'd never know that his colleagues argued about whether Jews in the suburbs should be allowed to drive to synagogue, or whether pushing a button on an elevator counted as work, or whether electricity was a form of fire that should not be lit. Avoiding work, for Heschel, did not mean being inactive. It meant avoiding the kinds of economic, "productive" busyness that occupied people six days of the week in order to create a space in which one could do other, more important things, and do them well. "Labor is a craft," he said, "but perfect rest is an art. To attain a degree of excellence in art, one must accept its discipline, one must adjure slothfulness." Heschel wasn't advocating passive *rest,* in other words, but *restoration.*

Those who get the most out of the digital Sabbath use it as a time to rebuild themselves, to reengage with friends, to relearn and exercise treasured predigital abilities, to reconnect with the real world. Turning off the million little requests and interactions that cascade into distraction and exhaustion is good, but trying to recover your mind just by unplugging is like trying to fix a building by abandoning it. The digital Sabbath is defined not only by what you turn off and ignore but also by what you do with the pauses. Unplugging is the means; rediscovering a more human sense of time and rebuilding your spirit are the ends.

EIGHT

EIGHT STEPS TO CONTEMPLATIVE COMPUTING

There are eight principles for contemplative computing. You're using them when you learn to be aware of how devices and media affect your breathing and mood; when you replace switch-tasking with real multitasking; when you adopt tools and practices designed to protect your attention; when you tweet mindfully; when you employ restorative spaces and digital Sabbaths to recharge your mind. Being familiar with the eight principles and seeing how they connect entanglement, Zenware, mindfulness, self-experimentation, and restoration can help you create relationships with information technologies that improve your extended mind. Their presence signals that you're using technologies in ways that will let you strengthen your mind and restore your focus and concentration; their absence is a sign that your relationship with your technologies isn't working for you.

The first principle is ***be human.*** In today's high-tech world, this means doing two things.

First, it means appreciating that entanglement is a big part of each of us. Humans have a terrific capacity to use technologies so well that they become invisible; we incorporate them into our body schema, employ them to extend mental and physical abilities. This is an ability our species has honed over more than a million years; the evolution of hands and invention of tools, the conquest of fire, the domestication of plants and animals for food and clothing, the invention of language and writing—all have made us both more human and more deeply entangled with technologies. We shouldn't resist entanglement with information technology. We should recognize the stakes, demand the chance to do it well, and insist on devices that serve and deserve us. (If there is ever a Cyborg Bill of Rights, that should be its first amendment.)

Second, it means recognizing how computers affect the way we see ourselves. Information technologies are evolving before our eyes, vastly increasing in power and sophistication, invading every corner of our lives, and they seem poised to match and exceed our own intelligence. Every year, our caveman-era brains feel less suited to the high-tech world. And even ordinary interactions with computers can do a great job of making us feel stupid. As a result, it's easy to see our own brains as puny and slow-moving and to feel a sense of resignation over our approaching cognitive obsolescence as our new computer overlords surpass human intelligence and memory. But remember that human intelligence and memory are different than their digital equivalents. We use the same words for them, but that obscures their very great differences. Remember that real time is not human time but the expression of a belief that the speed of commercial and financial transactions can be ratcheted ever upward; that the lag between events and reporting on events can be reduced to zero; and that people have to take less time to read, decide, and respond to changes in the world and workplace—none of which has to be true.

The second principle is ***be calm.*** The Calming Technology Laboratory's effort to create tools to promote "restful alertness" meshes perfectly with ancient ideas about calm as a foundation for contemplation.

We often think of calm as a purely physical state, the absence of disturbances in our minds or surroundings. People are calm when they're lying on the beach during vacation, away from the office and everyday cares. Contemplation, though, aims for a different kind of calm; it's active rather than passive, disciplined and self-aware. It's the deadly placidity of the samurai, the cool under pressure of an experienced pilot, the product of a masterful engagement that fills one's attention and leaves no room for distraction.

This type of calm requires training and discipline and a deep understanding of both devices and self. But it's not a calm that requires getting away from the world; it allows for fluid, quick action in the world. The aim isn't to escape, but to engage—in our case, to set the stage on which we can bring our entanglement with devices and media under our control so we can more effectively engage with the world and extend ourselves.

The third principle is ***be mindful.*** Learn what being mindful feels like, and learn to see opportunities to exercise it while online or using devices.

One reason meditation is such a valuable tool for contemplative computing is that it provides a simple and unadorned experience of mindfulness. You can be mindful doing everything from archery to motorcycle maintenance, but because such activities offer a wide variety of challenges and satisfactions, it can be hard to identify those parts that engage your concentration and facility for unobstructive self-observation. By stripping experience down to a bare minimum and giving the mind nothing to focus on but itself, meditation increases your ability to recognize mindfulness when you experience it in the wild.

Buddhist monks and nuns see being online as a chance to practice mindfulness. I can't think of a place less calming than the Internet, but Sister Gryphon explained it nicely. Even after years as a nun, she says, she sometimes still gets distracted. "One minute I'm watching a teaching by Chögyam Trungpa Rinpoche, the next minute I'm watching a video of a cat that barks like a dog," she says with a laugh. "But is that bad? We are living, curious little beings." Her Zen teachers urged her not to be discouraged when she noticed her mind wandering during meditation. It's like lifting a dumbbell, they would say; every time you return your mind's focus, your meditation gets stronger. Like her fellow monastics, Gryphon sees the Web as an arena of distraction where she will be challenged to stay mindful and speak and act compassionately.

Buddhist monastics treat the Web as a place to test their mindfulness, capacity for compassion, and right behavior. The digital world's distractions and impersonality make it easy to forget that you're ultimately interacting with real people, not just Web pages. Damchoe Wangmo recommends you "investigate your motivation before each online action, to observe what is going on in your mind," and stop if you're driven by "afflictive emotions" like jealousy, anger, hatred, or fear. Choekyi Libby watches herself online to "make sure I'm doing what I'm doing motivated by beneficial intention." As Marguerite Manteau-Rao and Elizabeth Drescher argue, the task with contemplative computing is not to make technology increase your empathy—even great design can't eliminate bad behavior—but to bring empathy to technology, to have your interactions be informed by your own ethical guidelines and moral sensibility. If you can be a positive presence online, you can be an even better one in the real world.

Approaching your interactions with information technologies as opportunities to test and strengthen your ability to be mindful; treating failures to keep focused as normal, predictable events that you can learn from; observing what helps you be mindful online and what doesn't—in other words, engaging in self-observation and self-experimentation—can improve your interactions with technologies and build your extended mind.

Mindfulness supports the fourth principle of contemplative computing, which is ***make conscious choices.*** Perhaps more effectively than any other technology in history, computers do a fabulous job of making themselves look invincible and inevitable; they're too powerful, too pervasive, too intrusive, too much fun, too useful to avoid. But that doesn't mean you have to surrender to them. Rather, you can live with information technologies while still being rightly jealous of your attention and freedom—conserving where possible, trading them only for things of greater value.

You can make more thoughtful, deliberate decisions about what technologies to use and how to use them if you know your aims, your tools, and your mind. The ability to self-observe allows readers choosing between print books and e-readers to know their aims and see how media affordances support the way they need to read.

Sometimes mindful choices involve exchanging old skills for new. In architecture, adopting CAD gave designers new abilities—to simulate energy use and airflow, to collaborate more effectively with engineers and construction foremen, and to experiment with new styles of building—but at the cost of abandoning a long tradition of drawing and, along with it, the rigor and thoughtfulness forced by its constraints and the embodied sense of visual proportion that came with constant practice.

Technologies meant to support mindfulness don't mislead you into thinking that devices limit your capacity to make choices or that you do not need to be responsible for your decisions. They remind you that you have free will. Zenware helps you focus your attention and reminds you that you make your own choices about where to direct it. As Fred Stutzman observes, having to restart your computer to disable Freedom puts you in the uncomfortable position of having "a moment of reflection... about why things have failed."

The fifth principle of contemplative computing is use devices in ways that ***extend your abilities.*** Technologies can amplify your natural faculties and senses, give you entirely new ones, and expand your extended mind—or they can be used as crutches, eroding cognitive abilities and weakening your mind.

Using devices in ways that extend our abilities means using them as tools for training and enriching our minds. Seeing the world through a camera lens has improved my visual attention; I notice colors, textures, the play of light on hard and dull objects, the three-dimensionality of wood grain and sea foam in a way I didn't before. Geotagging those photographs enriches my sense of a new place; relying on a GPS navigation system to guide me would not (and it would leave me vulnerable if its directions were faulty). Many users of Freedom discover that while it makes the Internet usefully inaccessible, it also helps them see that it's possible to curb self-distraction. Zenware blocks out distractions and externalizes a commitment to being more focused and creative. It doesn't substitute for self-discipline. It supports self-discipline.

Monastics speak of the need to view technology as a tool for solving one's problems, not as the solution itself. Sister Gryphon advises that while Zenware is useful, "Ultimately we must build up our own will power. Only we can be responsible for ourselves and what we are doing." YouTube will still be there when you go back online, and fighting the urge to self-distract leads to more distraction. "The true answer" to the problem of digital distraction "comes with seeing clearly and understanding ourselves and our reality. Then, there no longer is a struggle. What we want, how we wish to live, and how we are living will become one and the same."

These examples also show that you don't always face a choice between smarter tools *or* smarter selves, richer transactional memories *or* deeper and more reliable recall, photographic records *or* photographic memories. Contemplative computing often allows you to develop both.

The sixth principle of contemplative computing is ***seek flow.*** Flow, you'll recall, is that state that comes when you're completely absorbed in an activity. Your abilities and the challenge are perfectly balanced; the task is hard enough to be engaging but not hard enough to be discouraging. The world narrows; your attention filters out everything except the clues in the mystery, the twists in the road, the game board, the music score, the rock outcropping, the line of code, or the pattern in the data. Time seems to distort; you look up and realize that hours have passed without your noticing.

It's an immensely satisfying feeling, and it can be a great source of mental strength and psychological resilience. But those benefits are not guaranteed. Video games and Web browsing deliver flowlike experiences, but too often what you do onscreen doesn't help you in the real world. Game designers and Web developers have avidly read Csikszentmihalyi's *Flow,* and they're interested in the technical properties of flow—the details of flow experiences, how to encourage flow in users—but not in its larger uses.

You can move beyond the limited view of game and Web designers if you're mindful of the benefits of flow and the potential for turning all kinds of online and real-world experiences into flow. Buddhist monastics who treat the Web as a test of mindfulness are constructing a simple flow game for themselves; just as Mihaly Csikszentmihalyi's lox cutter made a game out of cutting the thinnest, largest number of slices from each salmon, monastics aim to be online without being distracted. Both activities sound absurdly simple, and they are, but they're more like the Japanese game of Go than like tic-tac-toe, for they share a simplicity that creates open-ended challenges, not boredom.

I ask Csikszentmihalyi if he was surprised that flow promoted psychological resilience and seemed to offer a key to living a good life. Was it odd to start out doing scientific research and end up doing moral philosophy? No, he replies, for him the field was never just "an appendage to counseling, or about how rats move through mazes." His interest in psychology, Csikszentmihalyi explains, dates from his childhood. He discovered the science as a teenager when he heard Carl Jung lecture in Switzerland (he had gone there on vacation, but there was too little snow to ski). Even before that, though, he was asking questions about how to live a good life. His father was a Hungarian diplomat, and Mihaly was born in Fiume, Italy, in 1934, where his father, Albert, ran the consulate. They later moved to Rome when Albert was appointed ambassador. For most of World War II, Csikszentmihalyi recalls, they were reasonably comfortable. As the war came to its chaotic end, "in the fall of forty-four, everything started going awry." One older brother was killed, and another disappeared into the Soviet gulags, where he spent six years. Csikszentmihalyi himself spent time in an Italian prison camp. "At that point, I suddenly realized that all these adults I was looking up to, thinking they had the keys to an understanding of life, were completely clueless," he recalls. "That experience of those months at the end of the war when everything went kaput suggested to me that you had to find a better way of living."

Things went further downhill after the war. When the Communists came to power in Hungary, they attacked the aristocracy, stripped them of their property, barred them from higher education, and forced most of them into exile. Many of his family's friends, high-level academics and civil servants, lost everything and became "zombie-like." They couldn't cope. The Csikszentmihalyis didn't have it any easier. His father despised the Communists and resigned his ambassadorship rather than cooperate with the regime. Overnight, they went from being members of Rome's diplomatic elite to being refugees.

But rather than give up, his father "sold some paintings and did what he wanted to do all along, which was open a restaurant.... Most people didn't have that kind of resilience," Csikszentmihalyi notes wryly. Albert soon discovered, incredibly, that "he liked serving food better than being an ambassador." For years, the family business was "the most chic restaurant in Rome, two minutes from the Trevi fountain." Bogart and Bacall dined there when they were in town. Mihaly waited tables.

Most former ambassadors would consider going into the restaurant business a humiliation, but Albert's sons offer some clues as to why their father found it rewarding. During the war, the young Csikszentmihalyi discovered an ability to lose himself for hours in playing chess or painting. His older brother, an expert in geology, could spend a day absorbed by the challenge of reconstructing the history of a rock sample. Their shared capacity for deep mental absorption gave them pleasure in the moment and helped them be resilient and adaptable in the face of life-shaking events. Perhaps their father found the restaurant business just as absorbing. For someone with this kind of mental ability, the thousand challenges of getting together the money, navigating the Italian bureaucracy, and finding a location, and then the endless daily work of putting together a menu, dealing with customers, and making each meal as perfect as possible would have made for a great life.

In other words, in the war and its aftermath, in his own experience and in those of his family, Csikszentmihalyi had found his better way of living, the thing that could give life meaning even during the hardest times. He had discovered flow.

So it's no surprise that *Flow* is concerned with issues of happiness, resiliency, and the foundations of a good life. They weren't incidental to the project. Understanding flow was always a means to an end. For Csikszentmihalyi, it was about how to be happy, how to keep your world together when everything is threatened, how to marshal the resources and resilience to remake yourself when your old life becomes unsustainable.

The seventh principle of contemplative computing is use technologies in ways that ***engage you with the world.***

This happens naturally when you use information technologies so fluently that your awareness of them falls away, and they become effectively invisible to you. When they cease to require your conscious attention, when they become part of your extended self, they can make you more aware of the world—the physical world, the worlds of others, the world of ideas.

Engagement with the world improves when you back away from activities that divide your attention. If live-tweeting requires you to shift your attention from interesting events to the work of updating, and if taking pictures or video means you're fussing with the device more than you're watching the moment, then you should avoid them. However, if you can use them to be more in the moment, to see more clearly, or to listen more closely, then go for it. Thomas Merton's contemplative approach to photography allowed him to treat the camera as a tool to sharpen his vision, and that enlarged his capacity to observe the world. Some people say live-tweeting makes them listen more carefully at lectures and conferences. Personally, I prefer to take notes during events—to stay attentive by writing, just like my tweeting colleagues—and publish something later after I've had time to reflect. You need to experiment and figure out what works best for you.

Engaging with the social world isn't just about interacting; it's about interacting constructively and ethically. It's about putting people rather than technology at the center of your attention. For some, this involves applying Christian or Buddhist precepts to their virtual interactions and using media in ways that let them be spiritual presences, not just social ones, and see the spark of divinity in everyone.

Engagement with ideas often involves the kind of vanishing act that Zenware is designed to perform. Even design mavens who admire the Dieter Rams–style minimalism that Jesse Grosjean brought to WriteRoom are supposed to be able to look past that and focus on the words and ideas. Typographers have long said that great type is like a wine goblet: you may appreciate the goblet's delicate lines and beautiful clarity, but you shouldn't be able to taste it. The best tools are ones that vanish when you no longer pay attention to them, that become part of your extended mind and elaborated body schema.

The eighth principle of contemplative computing is use—or abstain from—technologies in ways that are ***restorative,*** that renew your capacity for attention.

Attention and focus aren't always easy to direct. Often you have to work to keep your attention from wandering away from a task or screen or project. Concentration doesn't naturally bubble up in the absence of distractions; the resting mind does a great job of distracting itself. There are limits to how long we can focus deeply on anything (or, in the case of meditation, on nothing at all). Concentration is like strength: it can be developed with practice, but it's depleted with overuse and needs to be renewed.

So knowing how to restore the mind's ability to focus is essential. You can arrange your environment—from your screen environment to your immediate physical environment—and make it easier to concentrate for longer periods. It's essential to find activities that offer a respite but not a complete break from steady concentration. Things that offer a mix of fascination, a sense of your being away, boundlessness, and compatibility are most likely to let the conscious mind recharge.

Practicing this kind of restoration is especially important when you're working on complex problems that take weeks or months to solve and that require enormous amounts of intellectual energy. It's critical to let the mind be diverted enough to be restored, relieving the conscious mind while letting the subconscious keep working.

Let's end where we began: on the western edge of the ancient city of Kyoto, Japan, in the shadow of the stormy mountain Mount Arashiyama.

Iwatayama Monkey Park is not the only point of interest in the area. At the base of Arashiyama, below the monkey park, is the Tenryū-ji Temple, a Zen Buddhist complex. Even in a city filled with ancient treasures, it is highly venerated. Founded in 1339 and built on the site of a much older Zen temple, the Temple of the Heavenly Dragon (as *Tenryū-ji* is translated) at one time included teaching and meditation halls, residences for the abbot and monks, kitchens, and numerous small temples, some hundred and fifty buildings in all. The Zen practiced there was famous for its rigor and austerity, and in its early years, it had a strong influence on the culture of the samurai.

Today, only a few of those buildings remain, but the site is still spectacular thanks to the garden laid out by the temple's first abbot, the renowned Zen master Musō Sōseki. From the veranda of the main hall, one looks out across a pond to a *kasan,* an artificial mountain range. A broad walking path hugs the edge of the pond, continues through the garden, winds through a green bamboo forest, and finally opens onto a dry garden. (Darwin would have loved it.)

Musō Sōseki was not just Tenryū-ji's founding abbot. Musō was the Steve Jobs of early Zen, a serial entrepreneur with a tremendous design sense. Tenryū-ji was the sixth and grandest temple Musō founded, and its garden is one of his most famous. He pioneered the dry rock garden, an austere microcosm of a landscape. He integrated walking paths into his gardens, turning them from spaces one observed at a distance to places one experienced. But even more important, he integrated gardens into Zen teaching and practice. "He who distinguishes between the garden and practice," Musō said, "cannot be said to have found the true Way." Monks at Tenryū-ji still refer to the garden as a teacher.

Zen Buddhism holds that satori, or enlightenment, is not to be found by analyzing texts but through meditation, through calming the body and analyzing the mind. The garden was not designed to provide a respite from monastic practice. It was designed as a space to inspire and guide contemplation. It also serves as a reminder that body and mind are not separate entities; satori cannot be found unless both are involved. Musō's Zen garden is a technology based on the ideas that body and mind are inseparable, that satori is an active state, and that a well-designed garden can support contemplation by exploiting one's deep natural capacity to become entangled with spaces and tools. Even if you're in the shadow of monkeys, contemplation is within reach.

One of the wisest Buddhist sayings is that pain is inevitable, but suffering is a choice. Loss and death are unavoidable. Friends come and go, loved ones die, catastrophes strike, and eventually we must all come to terms with our own mortality. It's not within our power to escape these things, but we can develop the capacity to deal gracefully with them. We can learn from painful experiences, become wiser and better through them—and make ourselves better prepared for the next setback.

You face a similar situation in this superconnected, high-tech world. Information technologies are inescapable. They're part of how you work, how you keep in touch, how your kids play, how you think and remember. They clamor for your time and crave your attention. They rely on the fact that your relationships with information technologies are deep and profound and reflect the entanglement with tools that defines us as a species. They promise to be helpful and supportive, to make you smarter and more efficient, but too often they leave you feeling busier, distracted, and dull. Some say that the unavoidable price of being always on and connected is that one's attention is perpetually fractured, the mind subject to endless demands and distractions. But that's wrong. You are the inheritor of a contemplative legacy that you can use to retake control of your technologies, to tame the monkey mind, and to redesign your extended mind. Connection is inevitable. Distraction is a choice.

The Sandwalk today (courtesy of English Heritage)

APPENDIX ONE

KEEPING A TECH DIARY

This is a version of the technology diary that Ohio State University professor Jesse Fox assigns to her students. It's meant to give you a more detailed sense of the kinds of things you should track when gathering data for your self-experiments. Jesse's generosity in sharing this and allowing it to be reprinted is a model of good scholarly behavior and an example of how social scientists can (and should) share their insights with broader audiences.

1. For one typical weekday and one typical weekend day, keep a notebook (or note function on your technology of choice) and keep track of every mediated/technological interaction you have over the course of the day; note why you used the technology and how much time you spent with it. Write down the time you started and the time you stopped each interaction. Accessing Facebook? Write it down. Getting or sending a text? Write it down. Snapping a pic with your phone? Write it down. Checking e-mail? Write it down. Reading a PDF for class you downloaded? Write it down. Using your GPS? Write it down.

Listening to your iPod? Write it down. Watching a show on Netflix? Write it down. Playing a video or mobile game? Write it down. Be wary of times when you are using multiple technologies at once!

It might also be interesting for you to mark down how you spend the remainder of your time—sleeping, talking to friends face-to-face, studying without technology, using "old" media like magazines, books, or newspapers—to compare the proportion of your day spent with and without technology, to contrast old versus new media use, or to evaluate how much of your daily social interaction comes via technology.

2. Crunch your numbers. Report how much time total or how many times you used specific technologies. Where does your time go? What percentage of your waking hours is spent with technology?

3. Evaluate and criticize the tasks you achieved with your media use in light of the amount of time you spent. For each, consider:

(a) Do you feel this task required technology, was improved by using technology, or lost something because it was via technology?
(b) Did you have an emotional response to using the technology (e.g., relief, joy, frustration)? You should specify whether your emotional response was from the content (e.g., a mean text message from a friend), the technology itself (e.g., annoyance because the text message sound interrupted a conversation), or both.
(c) Were you multitasking? Do you feel your multitasking was effective?
(d) Was this a positive or negative experience? Do you think this was the optimal use of your time and the optimal use of the technology, or could you have done something better or more effectively?

4. Reflect. Considering your current technology use, what changes (if any) could you make to improve your day-to-day life (e.g., work productivity, mood, study habits, sleep, social interaction, health)? Are these goals compatible? What barriers exist to your making these changes?

As Jesse notes, "I leave it up to them to define *technology.* Although I give examples (mostly to remind them of all the small things they do that involve tech), I am always interested to see how many of them include, for example, their alarm clocks or a computerized vending machine. This usually opens up some interesting discussion in class about how we grow to take technologies for granted and what we imagine our future will be with modern technologies as common as a microwave or as hip as a cassette player."

APPENDIX TWO

RULES FOR MINDFUL SOCIAL MEDIA

Sometimes it seems that the Internet is designed to inspire bad behavior. The anonymity of message boards and comment systems encourages trolling; the rapid pace of social media makes people respond without thinking; text messaging leads inevitably to oversharing irrelevant or embarrassing things. But poor design needn't determine poor behavior. If you're mindful, you can practice contemplative computing even on Facebook and Twitter. Here are some rules that thoughtful social-media users follow.

Engage with care. Think of social media as an opportunity to practice what the Buddhists call right speech, not as an opportunity to get away with being a troll.

Be mindful about your intentions. Ask yourself why you're going onto Facebook or Pinterest. Are you just bored? Angry? Is this a state of mind that you want to share?

Remember the people on the other side of the screen. It's easy to focus your attention on clicks and comments, but remember that you're ultimately dealing with people, not media.

Quality, not quantity. Do you have something you really want to share, something that's worth other people's attention? Then go ahead and share. But remember the aphorism carved into the side of the Scottish Parliament: *Say but little and say it well.*

Live first, tweet later. Make the following promise to yourself: I, [insert name here], will never again write the words *OMG I'm [doing X] and tweeting at the same time LOL.*

Be deliberate. Financial journalist and blogger Felix Salmon once lamented that most people believe that online content is not supposed to be *read* but *reacted to.* Just as you shouldn't let machines determine where you place your attention, you shouldn't let the words of others drive what you say in the public sphere. Being deliberate means you won't chatter mindlessly or feed trolls. You'll say but little and say it well.

APPENDIX THREE

DIY DIGITAL SABBATH

In interviewing people who observe regular digital Sabbaths, I noticed that, while every digital Sabbath is unique, everyone follows a similar process when creating it. As with all contemplative computing, people observe how their entanglements with technologies work, think about how those relationships can be improved, and then find practices that fit their lives and extend and restore their minds. Following these guidelines can help you develop the Sabbath that works best for you.

Set a regular time. Sabbaths should follow a regular, practical schedule. Weekend days are often best. Unless you're a farmer or construction worker, it's hard to unplug during the regular workweek. Schedule either a full twelve hours, from when you wake up until you go to bed, or a twenty-four-hour period.

Figure out what to turn off. Do this ahead of time. Technical rules—e.g., anything with a screen, anything with a keyboard—are the easiest to set and follow; just don't go overboard (the display on the coffeemaker doesn't count as a screen). Dedicated Sabbatarians also follow behavioral rules; they may have some devices they stay away from but others they consider acceptable (e.g., single-player video games are out, but games you play with others are okay; e-mail and social media are verboten, but streaming movies are all right; the iPad you use at the office stays in the drawer, but the Kindle can come out).

Don't talk about digital Sabbaths. It's not quite as bad as *Fight Club* ("The first rule of Fight Club is: you do not talk about Fight Club"), but until the habit becomes more common, don't feel the need to advertise. It can be very helpful to coordinate it with friends or do it with another family (the kids will have more fun complaining together), but unless you want to have to explain to people that you're really not becoming an antisocial Luddite, you might be happier preserving the Sabbath peace by not broadcasting it.

Fill the time with engaging activities. The digital Sabbath should be active; it shouldn't be just a chance to catch up on laundry and pay bills. Do something you don't normally do, something challenging, engaging, and aggressively analog. Get out in the world (I'll leave the decision about the GPS up to you); cook something complicated; teach the kids how to fly-fish; find that hip eight-hundred-page novel you started last month and start marching through it. (Of course, if doing laundry and paying bills are psychically rewarding for you, then by all means; you have my blessing.)

Be patient. Like all contemplative activities, digital Sabbaths require some effort; it takes time for you to get into the spirit of the thing and lose that shaky need to check your BlackBerry. You won't see dramatic benefits overnight; you might not see them in a month. Give yourself at least twelve weeks. Think of it this way: In a normal year, you might check your phone 12,376 times; I'm asking you to try checking your phone only 11,968 times this year, and then reevaluate. Rather than 720 hours online this year, see how 696 hours works for you. And you spend eleven days a year waiting for your computer to perform various tasks. Maybe spending twelve days having it wait for you will feel better.

Be open to the spiritual qualities of the Sabbath. For many of us, this is a bit of a challenge. But stepping away from the normal frantic whirl of work and the Web offers you a real chance to reflect on how life ought to be lived, or at least for you to concentrate more intently on its good parts. Take it. And don't worry about discovering that you actually want to give it all up, move off the grid, and raise goats. That doesn't really happen.

Enjoy your escape from "real time." Abraham Heschel's idea of Sabbath time being cut from a different piece of space-time fabric feels more relevant, and more welcome, than ever. The digital Sabbath is a chance to escape the "tyranny of things," particularly the tyranny of things that chirp, vibrate, tweet, and plead for your attention, all the while promising you it'll be worth it. It's a chance to escape the unreality of real time and rediscover how to live at your own pace. It'll be worth it. I promise.

ACKNOWLEDGMENTS

This book started as a sabbatical research project at Microsoft Research Cambridge, in England. Without the unexpected and amazing generosity of Richard Harper, director of the Socio-Digital Systems Group at MSC, this project would never have happened. The experience of strategizing over coffee in the MSC lounge, working through ideas during walks through Grantchester Meadows to the Orchard, and talking shop over pints at the Eagle and the Pickerel left an indelible imprint on my work. Sam Kinsley and his group, at the University of West Reading, and Yvonne Richards and her lab, at Open University, provided critical early interest in the project. I was also fortunate that my fellow Microsoft visitor Annie Gentes spent the next year at Stanford working on a project that overlapped in interesting ways with my own.

While at Microsoft Research Cambridge, I began writing about contemplative computing on my blog (http://www.contemplativecomputing.org). For me, a blog is a commonplace book, sounding board, and advertisement; it offers a chance for me to note interesting articles, think aloud about subjects that I might later explore in greater depth, and spread the word about upcoming talks. As a result, some of the posts formed the basis for sections of this book, but over the course of editing and revising, they have been substantially changed (and I hope improved).

I owe a great debt to the people who gave so much of their time to answer questions, sit for interviews, talk on Skype, or respond to my e-mails. Such generosity is humbling and gratifying. I thank Sun Joo (Grace) Ahn, Pia Aitken, James Anderson, John Bartol, Gwen Bell, Jan Bilik, Jeff Brosco, David Brownlee, Michael Chorost, Shay Colson, Marzban Cooper, Ruth Schwartz Cowan, Mihaly Csikszentmihalyi, Susana Darwin, André Delbecq, Anna Digabriele, Anne Dilenschneider, Jill Davis Doughtie, Elizabeth Drescher, Elizabeth Dunn, Nancy Etchemendy, Morley Feinstein, Jesse Fox, Jesse Grosjean, Michael Grothaus, Steve Herrod, Hal Hershfield, William Huchting, Cody Karutz, Mike Kuniavsky, Donald Latumahina, Ho John Lee, Chris Luebkeman, Lambros Malafouris, Marguerite Manteau-Rao, Neema Moraveji, Ramez Naam, Daniel Nevins, Colin Renfrew, Martha Rock, Christine Rosen, Prime Sarmiento, Sharon Sarmiento, Lauren Silver, Monica Smith, Linda Stone, Tammy Strobel, Fred Stutzman, Phil Tang, Edward Tenner, Mads Thimmer, Anthony Townsend, Lyn Wadley, Carolyn Wilson, David Wuertele, James Yu, and a pseudonymous contributor, Megan Jones. Special thanks to the Buddhist monks, nuns, and lay practitioners for their thoughtful and generous responses to my questions: Jonathan Coppola, Caine Das, Sister Gryphon, Choekyi Libby, Bhikkhu Samahita, Damchoe Wangmo, and Noah Yuttadhammo.

My literary agent Zoë Pagnamenta and my editor John Parsley have been essential collaborators. The rest of the Pagnamenta Agency—particularly the ever-patient Sarah Levitt—and everyone at Little, Brown have all been terrific. If you haven't written a book, you may imagine it as springing fully formed from the brow of the author. It never works that way. While I'm ultimately responsible for every word, this book is a collaborative production.

While the book began in Cambridge, it brings together a number of interests pursued at widely different times. My thinking about technology will, it seems, be forever influenced by my training under Rob Kohler, Riki Kuklick, and Tom Hughes in the Department of History and Sociology of Science at the University of Pennsylvania in the 1980s. I'm still trying to answer questions about creativity and technology that Tom Hughes posed in the very first class I took in college. Years later, Bob McHenry started me thinking about the cultural and cognitive impacts of new media when he hired me to guide the *Encyclopedia Britannica*'s editorial division in the 1990s. Finally, while a research director at the Institute for the Future in the 2000s, I discovered the intellectual rewards that come from applying abstract, scholarly ideas to concrete, real-world problems.

My children displayed great patience during the months I retreated to the garage to write, and even greater fortitude when my wife and I left them in California during my Cambridge sabbatical. You did really well, guys. Thanks.

Finally, my deepest thanks to my wife, Heather, who endures everything the spouse of a writer has to with understanding and grace. I couldn't do it without you.

NOTES

It used to be that endnotes were elaborate fortresses of citations, Lego castles of obscure references whose inaccessibility helped protect authors from criticism (as in, "Well, if you can't *find* the 1957 Croatian edition of the *Festschrift,* you certainly can't demonstrate that my argument is wrong"), or places for scholars to name-check friends, attack adversaries, settle scores, and look erudite (as in, "See Professor Smith's brilliant but unjustly overlooked takedown of Professor Jones's misguided thesis, published in the fourth edition of the *Festschrift* [Zagreb, 1957]"). In these endnotes, I expand on some points that are interesting but would have been distracting to discuss in the main text, and I provide an overview of some of the literature that informs this book.

Thanks to Google Books and to many scientists' habit of publishing electronic preprints of their articles, several of the scholarly articles and books mentioned here are accessible to any general reader with an Internet connection. And if you can't track down a work, many of the people cited here have published related works that are more easily accessible.

So let's get started.

Introduction: ***Two Monkeys***

The concept of the monkey mind is well known in Buddhism, but its origins are not clear. Tracing the origins of an idea in Buddhism can be surprisingly hard; ideas may be pondered for generations before they are written down, and scholars have to follow ideas as they move among Hindi, Chinese, Korean, and Japanese texts.

Japanese macaques are described in Naofumi Nakagawa, Masayuki Nakamichi, and Hideki Sugiura, eds., *The Japanese Macaques* (New York: Springer, 2010), and Jean-Baptiste Leca et al., eds., *The Monkeys of Stormy Mountain: 60 Years of Primatological Research on the Japanese Macaques of Arashiyama* (Cambridge: Cambridge University Press, 2011). There's a long history of monkeys serving as mirrors or foils for humans—how we see primates says a lot about how we see ourselves. Japanese thinkers have been pondering the macaque for centuries. As anthropologist Emiko Ohnuki-Tierney puts it, "No other nonhuman being in the Japanese universe has been as closely involved [as the macaque] in the Japanese people's deliberations" about what makes humans special. On monkeys in Japanese culture, see Emiko Ohnuki-Tierney, *The Monkey As Mirror: Symbolic Transformations in Japanese History and Ritual* (Princeton, NJ: Princeton University Press, 1987); quote is from "The Monkey As Self in Japanese Culture," in Ohnuki-Tierney, ed., *Culture Through Time* (Stanford: Stanford University Press, 1990), 129–30.

On the work of Miguel Nicolelis, see Nicolelis, *Beyond Boundaries: The New Neuroscience of Connecting Brains with Machines—and How It Will Change Our Lives* (New York: St. Martin's Press, 2012). Nicolelis's first major discovery came in 2001 when he implanted a set of electrodes in the brain of a monkey and attached the other ends to a robotic arm. The monkey could now, in theory, control the robotic arm with its brain—if it could learn how to do so. What Nicolelis wanted to understand was how the brain learned to master a new ability. Would it take a long time? Would it be successful at all? Would the monkey always have to think consciously about the arm, or would it eventually be able to control the arm as easily as it did one of its natural arms? Nicolelis and his team already knew a lot about BCI technologies, implants, and robotics in 2001. Two years earlier, his group had implanted electrodes in an owl monkey named Belle and, using juice as a reward, trained her to operate a computer joystick; as she moved the joystick around, a computer connected to the electrodes recorded which neurons in her brain fired and what actions they corresponded to. The scientists then disconnected her from the computer and hooked her up to two robotic arms—one at Duke and, just to make things a bit more spectacular, one at MIT, several hundred miles to the north. When Belle moved her joystick, the same signals that her brain sent to her arm now operated the robotic arms as well.

Belle had no idea that she was also operating robots. She couldn't see the robots and probably was more focused on the juice anyway. In the experiment in 2001, the new monkey had a joystick and also was able to see what the robotic arm was doing. As the monkey moved her joystick, the robot arm moved also, and those motions would control a cursor on a computer screen. Opening or closing the robot's hand made the cursor bigger or smaller. Not surprisingly, it took the monkey a little while to learn to use the arm and figure out the game. Once she mastered the game, the scientists turned off the joystick. The monkey now had only to think about the actions she needed to perform, and the robot would respond.

One more note on the term *contemplative computing.* A term like *pervasive computing* or *ubiquitous computing* describes a type of computer or information service that's been enabled by a dramatic new innovation. Personal computing, for example, developed when microprocessors and memory became cheap enough to produce computers that individuals, not just governments and big businesses, could afford. So when computer scientists and engineers talk about ubiquitous computing, proactive computing, engaged computing, Cloud computing, and other forms of computing, they're talking about new technologies and the innovations that make them possible.

The statistical information for the first Monday-morning scenario comes from a variety of sources.

Driving under the influence of electronics has received a lot of attention in the past few years. In a 2011 Unisys survey, nearly 50 percent of respondents said they had used mobile devices while in a car, and 20 percent admitted to using computers while in the driver's seat. See Klint Finley, "Always On: Your Employees Are Working and Driving," *ReadWrite* (July 12, 2011), online at http://www.readwriteweb.com/enterprise/2011/07/always-on.php. The most detailed analysis of technology-related distractions in police-cruiser driving is in Judd Citrowske et al., *Distracted Driving by Law Enforcement Officers Resulting in Auto Liability Claims: Identification of the Issues and Recommendations for Implementation of a Loss Control Program* (Saint Mary's University of Minnesota Schools of Graduate and Professional Programs, 2011), online at http://policedriving.com/wp-content/uploads/2011/10/Distracted-Driving-Saint-Marys-University-April-20111.pdf.

Gloria Mark and colleagues have found a correlation between frequent e-mail checking and increased stress; see Gloria J. Mark, Stephen Voida, and Armand V. Cardello, "'A Pace Not Dictated by Electrons': An Empirical

Study of Work Without Email," *Proceedings of the SIGCHI Conference on Human Factors in Computing Systems (CHI '12)* (Austin, Texas: May 5–10, 2012).

The Harris/Intel survey is described in Patrick Darling, "Stressed by Technology? You Are Not Alone," *Intel Newsroom Blog* (August 19, 2010), online at http://newsroom.intel.com/community/intel_newsroom/blog/2010/08/19/stressed-by-technology-you-are-not-alone.

Tel Aviv University researchers Tali Hatuka and Eran Toch's Smart-Spaces project (http://smartspaces.tau.ac.il/) is studying the impact of smartphones on situational awareness. Their work is described in "Smart Phones Are Changing Real World Privacy Settings," Tel Aviv University press release (May 12, 2012), online at http://www.aftau.org/site/News2?page=NewsArticle&id=16519.

Since 2008, sleep scientists have observed people sleep-texting. As sleep researcher David Cunnington explained, "Because it's so easy to receive emails constantly, and get notifications from smartphones, it becomes more difficult for us to separate our waking and sleeping lives"; Naomi Selvaratnam, "People Are Sending Text Messages While They Are Asleep, Says Specialist," *Herald Sun* (November 22, 2011), online at http://www.news.com.au/technology/texting-in-your-sleep-not-gr8/story-e6frfro0-1226201995575. See also Sandra Horowitz, "M-F-064, Sleep Texting: New Variations on an Old Theme," *Sleep Medicine* 12, supp. 1 (September 2011): S39.

Global and U.S. statistics on numbers of technological devices per home come from the International Telecommunications Union's report *Measuring the Information Society 2010* (Geneva: ITU, 2011). My household, which is admittedly a little more tech-intensive than many, has an electronic menagerie of one desktop computer, one Wii, one DVR, three laptops, three iPads, three Nintendo DS handhelds, four digital cameras, four cell phones, and about six iPods (no one is really sure anymore). That's an average of six devices per person. (You could also include a miscellany of backup drives, jump drives, DVD burners, the Newton MessagePad I can't seem to get rid of, and the household appliances that need to have their clocks reset twice a year.)

The estimate of 110 e-mails sent and received is from Quoc Hoang, "Email Statistics Report, 2011–2015," ed. Sara Radicati (Palo Alto, CA: Radicati Group). In a 2010 survey, 60 percent of Facebook users checked their pages five or more times per day; "Reader Redux: How Many Times a Day Do You Check Facebook?," *Geek Sugar* (March 25, 2010), online at http://

www.geeksugar.com/How-Many-Times-Do-You-Check-Facebook-One-Day-7891146. On phone checking, see Antti Oulasvirta et al., "Habits Make Smartphone Use More Pervasive," *Personal and Ubiquitous Computing* 16, no. 1 (January 2012): 105–14. Statistics on smartphone activity come from "Making Calls Has Become Fifth Most Frequent Use for a Smartphone for Newly Networked Generation of Users," *O2 News Centre* (June 29, 2012), online at http://mediacentre.o2.co.uk/Press-Releases/Making-calls-has-become-fifth-most-frequent-use-for-a-Smartphone-for-newly-networked-generation-of-users-390.aspx. In 2008, British research firm YouGov coined the term *nomophobia* ("no-mobile-phone phobia") to describe cell-phone-related anxiety. In a survey of 2,100 adults, they found that 53 percent felt anxious if they didn't have their cell phones or lost network coverage, and over 20 percent admitted they never turned their phones off; Robert Charette, "Do You Suffer from Nomophobia?" *IEEE Spectrum Risk Factor Blog* (May 22, 2012), online at http://spectrum.ieee.org/riskfactor/telecom/wireless/do-you-suffer-from-nomophobia?

On work addiction and busyness as a sign of importance, see Leslie Perlow, *Sleeping with Your Smartphone* (Cambridge, MA: Harvard Business Review Press, 2012). Multitasking's emotional satisfactions are studied in Zheng Wang and John M. Tchernev, "The 'Myth' of Media Multitasking: Reciprocal Dynamics of Media Multitasking, Personal Needs, and Gratifications," *Journal of Communication* 62, no. 3 (June 2012): 493–513.

On distraction and technology, see Maggie Jackson, *Distracted: The Erosion of Attention and the Coming Dark Age* (Amherst, NY: Prometheus, 2008); Jonathan B. Spira, *Overload! How Too Much Information Is Hazardous to Your Organization* (New York: John Wiley, 2011); Victor M. González and Gloria Mark, " 'Constant, Constant, Multi-Tasking Craziness': Managing Multiple Working Spheres," *CHI '04* (Vienna, Austria: April 24–29, 2004); Laura Dabbish, Gloria Mark, and Victor González, "Why Do I Keep Interrupting Myself?: Environment, Habit and Self-Interruption," *CHI '11* (Vancouver, BC: May 7–12, 2011).

Figures on device consumption, usage, and time spent online are drawn from Janna Anderson and Lee Rainie, *Millennials Will Benefit and Suffer Due to Their Hyperconnected Lives,* Pew Internet and American Life Project, 2012; U.S. Census, *2012 Statistical Abstract;* Aaron Smith, *Mobile Access 2010,* Pew Internet and American Life Project, 2010; United States Energy Information Administration, 2009.

The phrase *natural-born cyborgs* comes from Andy Clark's *Natural-Born Cyborgs: Minds, Technologies, and the Future of Human Intelligence* (Oxford: Oxford University Press, 2004). Clark, occupant of the wonderfully antique-sounding chair in Moral Philosophy at the University of Edinburgh, is one of today's most accessible yet rigorous writers exploring the philosophical implications of neuroscience and information technology.

Chapter 1: *Breathe*

On sleep apnea, see Terry Young, Paul E. Peppard, and Daniel J. Gottlieb, "Epidemiology of Obstructive Sleep Apnea: A Population Health Perspective," *American Journal of Respiratory and Critical Care Medicine* 165 (2002): 1217–39.

Andy Clark's *Supersizing the Mind: Embodiment, Action, and Cognitive Extension* (Oxford: Oxford University Press, 2010) reprints his and David Chalmers's "The Extended Mind," *Analysis* 58 (1998): 7–19, where they first presented the extended-mind thesis. Alva Noë's brilliant but technical *Action in Perception* (Cambridge, MA: MIT Press, 2006) and his more accessible *Out of Our Heads: Why You Are Not Your Brain, and Other Lessons from the Biology of Consciousness* (New York: Hill and Wang, 2010) are also outstanding.

Ironically, the high-tech use of the term *addiction* harks back to the word's ancient roots. The word *addiction* first appears in Shakespeare's *Henry V,* when one character says of young Prince Hal, "His addiction was to courses vain." The term comes from the Latin *addictus,* which was a kind of legal enslavement: as part of their punishment and as a way for them to make restitution, Roman debtors could be sentenced—*addicted*—to their creditors. (The law further stated that someone with multiple debts could be dismembered and his body divided among his several creditors. This tended not to happen.) The modern and more familiar use of *addiction* first appeared in the early 1900s, in reference to opium and morphine use. It made its way into high-tech thinking in the 1980s, when designers tried to create personal computers that were "easy enough for your grandmother to use." Before long, a product that was addicting was considered a good thing: it meant a regular user base and dependable revenues. So, in a way, an addictive social media resonates with the archaic and the modern meanings of the term: people who are addicted to Twitter are both enslaved to a compulsion that they cannot control and addicted to other users.

The University of Maryland study is *A Day Without Media,* online at http://withoutmedia.wordpress.com.

On tool use and human evolution, see Timothy Taylor, *The Artificial Ape: How Technology Changed the Course of Human Evolution* (London: Palgrave Macmillan, 2010); Stanley H. Ambrose, "Paleolithic Technology and Human Evolution," *Science* 291, no. 5509 (March 2, 2001): 1748–53; and Richard Wrangham, *Catching Fire: How Cooking Made Us Human* (New York: Basic Books, 2010). According to Australian archaeologist Thomas Suddendorf, the manufacture of stone tools in one area for use somewhere else is proof that "mental time travel" was a distinctive feature of protohuman consciousness; see Thomas Suddendorf, Donna Rose Addis, and Michael C. Corballis, "Mental Time Travel and the Shaping of the Human Mind," *Philosophical Transactions of the Royal Society, Biological Sciences* 364 (2009): 1317–24. Jane Hallos also makes the case that tool-making is evidence of planning ability in her "'15 Minutes of Fame': Exploring the Temporal Dimension of Middle Pleistocene Lithic Technology," *Journal of Human Evolution* 49 (2005): 155–79.

Experiments to teach bonobos to make tools were conducted at Indiana University in the 1990s; see Kathy D. Schick et al., "Continuing Investigations into the Stone Tool-Making and Tool-Using Capabilities of a Bonobo (*Pan paniscus*)," *Journal of Archaeological Science* 26, no. 7 (July 1999): 821–32.

Humans have worn clothes for at least the past 170,000 years, a date established by scientists' study of the evolution of body lice, which have evolved to live on clothed skin, not naked skin; see Melissa A. Toups et al., "Origin of Clothing Lice Indicates Early Clothing Use by Anatomically Modern Humans in Africa," *Molecular Biology and Evolution* 28, no. 1 (January 2011): 29–32. Shoes, by contrast, are a relatively recent innovation, and seem to be about 40,000 years old, first worn during the Upper Paleolithic transition; see Erik Trinkaus, "Anatomical Evidence for the Antiquity of Human Footwear Use," *Journal of Archaeological Science* 32, no. 10 (October 2005): 1515–26.

On drugs, see Richard Evans Schultes, Albert Hofmann, and Christian Rätsch, *Plants of the Gods: Their Sacred, Healing, and Hallucinogenic Powers,* rev. ed. (Rochester, VT: Healing Arts Press, 2001). Schultes was one of the founders of ethnopaleobotany, while Hofmann is best known as the discoverer of LSD.

The cognitive impacts of writing on Greek civilization and thinking are described in Eric Havelock's short, brilliant book *The Muse Learns to Write*

(New Haven: Yale University Press, 1986). Walter Ong, *Orality and Literacy: The Technologizing of the Word,* rev. ed. (1982; repr. London: Routledge, 2002), is also excellent.

The Mycenaean sword case is made in Lambros Malafouris, "Is It 'Me' or Is It 'Mine'? The Mycenaean Sword As a Body-Part," in J. Robb and D. Boric, eds., *Past Bodies* (Oxford: Oxbow Books, 2009), 115–23. Malafouris is a leading figure in cognitive archaeology; see, among other important works in the field, L. Malafouris, "The Cognitive Basis of Material Engagement: Where Brain, Body, and Culture Conflate," in E. DeMarrais, C. Gosden, and C. Renfrew, eds., *Rethinking Materiality: The Engagement of Mind with the Material World* (Cambridge: McDonald Institute for Archaeological Research, 2004), 53–62; L. Malafouris, "Beads for a Plastic Mind: The 'Blind Man's Stick' (BMS) Hypothesis and the Active Nature of Material Culture," *Cambridge Archaeological Journal* 18, no. 3 (2008): 401–14; "Between Brains, Bodies and Things: Tectonoetic Awareness and the Extended Self," *Philosophical Transactions of the Royal Society, Biological Sciences* 363 (2008): 1993–2002; and Dietrich Stout et al., "Neural Correlates of Early Stone Age Toolmaking: Technology, Language, and Cognition in Human Evolution," *Philosophical Transactions of the Royal Society, Biological Sciences* 363 (2008): 1939–49. The cognitive archaeology of weapons is described in Marlize Lombard and Miriam Noël Haidle, "Thinking a Bow-and-Arrow Set: Cognitive Implications of Middle Stone Age Bow and Stone-Tipped Arrow Technology," *Cambridge Archaeological Journal* 22, no. 2 (2012): 237–64.

On body schema, see Lucilla Cardinali, Claudio Brozzoli, and Alessandro Farnè, "Peripersonal Space and Body Schema: Two Labels for the Same Concept?," *Brain Topography: A Journal of Cerebral Function and Dynamics* 21, no. 3–4 (2009): 252–60.

The curious phenomenon of phantom cell-phone vibration is documented in David Laramie, "Emotional and Behavioral Aspects of Mobile Phone Use" (PhD diss., Alliant University International, 2007); Ghassan Thabit Saaid Al-Ani, Najeeb Hassan Mohammed, and Affan Ezzat Hassan, "Evaluation of the Sensation of Hearing False Mobile Sounds (Phantom Ring Tone; Ringxiety) in Individuals," *Iraqi Postgraduate Medical Journal* 1, no. 1 (2009): 90–94; Michael Rothberg et al., "Phantom Vibration Syndrome Among Medical Staff: A Cross-Sectional Survey," *British Medical Journal* 341 (2010): c6914; Michelle Drouin, Daren H. Kaiser, and Daniel A.

Miller, "Phantom Vibrations Among Undergraduates: Prevalence and Associated Psychological Characteristics," *Computers in Human Behavior* 28, no. 4 (July 2012): 1490–96.

The description of the airplane as a "beautiful machine" is from NASA astronaut David Randolph Scott, in David Scott and Alexei Leonov, *Two Sides of the Moon: Our Story of the Cold War Space Race* (New York: St. Martin's, 2006), 29.

Ellen Ullman's wonderful *Close to the Machine: Technophilia and Its Discontents* (London: Picador, 2012) is a great inside look at the seductions of programming.

On the embodied nature of mathematics, see George Lakoff and Rafael Núñez, *Where Mathematics Comes From: How the Embodied Mind Brings Mathematics into Being* (New York: Basic Books, 2000). There's also an interesting literature (if you're into this sort of thing) analyzing the gestures mathematicians and math teachers use to explain concepts as evidence of the embodied nature of mathematics; see Martha W. Alibali and Mitchell J. Nathan, "Embodiment in Mathematics Teaching and Learning: Evidence from Learners' and Teachers' Gestures," *Journal of the Learning Sciences* 21, no. 2 (2012): 247–86, and Nathalie Sinclair and Shiva Gol Tabaghi, "Drawing Space: Mathematicians' Kinetic Conceptions of Eigenvectors," *Educational Studies in Mathematics* 74, no. 3 (2010): 223–40.

The Greek typesetter story comes from John Seely Brown and Paul Duguid, *The Social Life of Information* (Cambridge, MA: Harvard Business School Press, 2000). The story of Otto is from Clark and Chalmers, "The Extended Mind."

Betsy Sparrow's work on transactive memory is described in Sparrow, Jenny Liu, and Daniel M. Wegner, "Google Effects on Memory: Cognitive Consequences of Having Information at Our Fingertips," *Science* 333, no. 6043 (August 5, 2011): 776–78.

On reading, underlining, and note-taking, start with Maryanne Wolf, *Proust and the Squid: The Story and Science of the Reading Brain* (New York: Harper Perennial, 2008). Then move on to more specialized studies: Sarah E. Peterson, "The Cognitive Functions of Underlining as a Study Technique," *Reading Research and Instruction* 31, no. 2 (1991): 49–56; Rebecca Sandak et al., "The Neurobiological Basis of Skilled and Impaired Reading: Recent Findings and New Directions," *Scientific Studies of Reading* 8, no. 3 (2004): 273–92; and Fabio Richlan, Martin Kronbichler, and Heinz Wimmer, "Functional Abnormalities in the Dyslexic Brain: A Quantitative Meta-Analysis of Neuroimaging Studies," *Human Brain Mapping* 30 (2009): 3299–3308.

The history of word spacing is told by Paul Saenger in two works: "Silent Reading: Its Impact on Late Medieval Script and Society," *Viator: Medieval and Renaissance Studies* 13 (1982): 367–414, and *Space Between Words: The Origins of Silent Reading* (Stanford: Stanford University Press, 1997).

Legal reading is described in Ruth McKinney, *Reading Like a Lawyer* (Durham, NC: Carolina Academic Press, 2005); Kirk Junker, "What Is Reading in the Practice of Law?" *Journal of Law and Society* 9 (2008): 111–62; Leah M. Christensen, "The Paradox of Legal Expertise: A Study of Experts and Novices Reading the Law," *Brigham Young University Education and Law Journal* 1 (2008): 53–87.

The classic work on flow is Mihaly Csikszentmihalyi, *Flow: The Psychology of Optimal Experience,* rev. ed. (1992; London: Rider, 2002). See also Mihaly Csikszentmihalyi and Isabella Selega Csikszentmihalyi, eds., *Optimal Experience: Psychological Studies of Flow in Consciousness* (Cambridge: Cambridge University Press, 1988). An accessible introduction to the science of attention is Winfried Gallagher, *Rapt: Attention and the Focused Life* (New York: Penguin, 2010).

A capacity to reach flow, to see challenges as things to be embraced rather than avoided, leads to a greater degree of resilience, the ability to deal with life-changing challenges. In his fascinating book *Deep Survival: Who Lives, Who Dies, and Why* (New York: W. W. Norton, 2004), Laurence Gonzales notes that people who survive being lost at sea, buried in an avalanche, having their neighborhoods destroyed by earthquakes or hurricanes, or other disasters share a few psychological traits. They're able to adjust quickly to the new reality, to accept the idea that help might not arrive for days and that they ultimately might not make it. At the same time, they're able to identify immediate tasks to keep themselves busy, to find patterns that bring mental order to the chaos of a suddenly broken universe, and to push thoughts of death out of their minds. (Interestingly, having a companion who's injured but not in mortal danger can increase your chances of survival, because caring for someone else forces you to focus and stop feeling sorry for yourself.) Their senses are sharpened, and they're able to respond to opportunities, but they can also find beauty in their situation. People who survive days in a lifeboat without water, for example, maintain the wherewithal to rig a tarp to catch rainwater but can also be captivated by a clear night sky or the glow of luminescent sea creatures. As one observer put it, "Survival is nothing more than an ordinary life well lived in extreme circumstances."

Eugen Herrigel's *Zen in the Art of Archery: Training the Mind and Body to Become One* (New York: Penguin, 2004) first appeared in English in 1953; it is still regarded as a classic explication of Zen, though in recent years it has been the subject of pointed criticism. It was based on a set of lectures Herrigel delivered on the Chivalrous Art of Archery in Berlin in 1936; Herrigel joined the Nazi Party the following year and enjoyed a successful academic career at the University of Erlangen, culminating with his appointment as rector in 1944. For a critical analysis of Herrigel's understanding of Japanese archery, his work, and the work's influence, see Yamada Shōji, "The Myth of Zen in the Art of Archery," *Japanese Journal of Religious Studies* 28 (2001): 1–30, and Yamada Shōji, *Shots in the Dark: Japan, Zen, and the West* (Chicago: University of Chicago Press, 2009).

Neema Moraveji's work is documented on the Calming Technology blog at http://calmingtechnology.org/; the Calming Coach is described in Neema Moraveji, "Augmented Self-Regulation" (PhD diss., Stanford University, 2012).

Chapter 2: *Simplify*

Unless otherwise noted, quotations in this chapter come from interviews with James Anderson, Marzban Cooper, Jesse Grosjean, Michael Grothaus, Rebecca Krinke, Donald Latumahina, and Fred Stutzman that I conducted in the summer and fall of 2011.

Freedom is available from Fred Stutzman's Web site at http://macfreedom.com/. WriteRoom is available at http://www.hogbaysoftware.com/products/writeroom. Virginia Heffernan talks about WriteRoom in "An Interface of One's Own," *New York Times* (January 6, 2008).

The perils of aircraft autopilot and fly-by-wire systems have been discussed for several years, particularly in the wake of the crash of Air France flight 447. In that case, the Airbus 330—one of the most sophisticated planes flying today—crashed after its autopilot failed and the copilot caused the plane to stall by pulling up on the controls. A number of safety experts and pilots noted that high-tech aircraft like Airbuses are easy to fly under normal circumstances, but it's very hard to make sense of what's happening when things go wrong, like an easy-to-use computer that shows you nothing but a frozen screen when it encounters a problem. The Airbus's complexity keeps pilots from developing the instincts they need to keep disabled planes in the air. Flight 447's last moments and the role pilot error played in the crash are described in Jeff Wise, "What Really Happened Aboard Air France 447," *Popular Mechanics* (December 6, 2011), online at http://www.popularmechanics.com/technology/aviation/crashes/what-really-happened-aboard-air-france-447-6611877.

On the history of multitasking, see Lyn Wadley, Tamaryn Hodgskiss, and Michael Grant, "Implications for Complex Cognition from the Hafting of Tools with Compound Adhesives in the Middle Stone Age, South Africa," *Proceedings of the National Academy of Sciences* 106, no. 24 (June 16, 2009): 9590–94; Monica Smith, *A Prehistory of Ordinary People* (Phoenix: University of Arizona Press, 2010).

Clifford Nass's findings on compulsive switch-tasking are described in Eyal Ophir, Clifford Nass, and Anthony D. Wagner, "Cognitive Control in Media Multitaskers," *Proceedings of the National Academy of Sciences* 106, no. 37 (September 15, 2009): 15583–87; see also Nass's interview with PBS's *Frontline,* December 1, 2009, online at http://www.pbs.org/wgbh/pages/frontline/digitalnation/interviews/nass.html, the source of the quote. For more on the costs of multitasking, see Nicholas Carr, *The Shallows: What the Internet Is Doing to Our Brains* (New York: W. W. Norton, 2010).

The idea of opera as multitasking is inspired by scholarly comparisons of opera and virtual reality, an idea first articulated in Michael Heim, *The Metaphysics of Virtual Reality* (Oxford: Oxford University Press, 1993), then picked up in Randall Packer and Ken Jordan's collection *Multimedia: From Wagner to Virtual Reality* (New York: W. W. Norton, 2001), and Matthew Wilson Smith, *The Total Work of Art: From Bayreuth to Cyberspace* (New York: Routledge, 2007).

The idea that creativity is driven by interesting juxtaposition is advanced by a number of writers, perhaps most cogently Silvano Arieti in his *Creativity: The Magic Synthesis* (New York: Basic Books, 1976). Examples of creative combinations can be found in every inventive field: Christopher Wren blended Baroque and Classical architecture in his design for St. Paul's Cathedral; George de Mestral used hooks on plant burrs as a model for a new class of fastener (now called Velcro); LA chef Roy Choi combined *bulgogi* and tortillas to make Korean tacos.

Creativity can flourish under confined but not desperate circumstances: Machiavelli, for example, was essentially under house arrest when he wrote *The Prince;* de Sade's wife smuggled writing materials and sex toys into the Bastille for her husband; and Braudel wrote his great work when he was in a

German prisoner-of-war camp for officers, which offered better conditions than other German camps, outside Lübeck. Prisoners and prisoners of war enduring harsh regimens or heavy labor aren't so lucky. And, of course, we all know who *else* wrote in prison: during his year in Landsberg Prison outside Munich, Adolf Hitler wrote *Mein Kampf*.

Jeffrey MacIntyre coined the term *Zenware* in "The Tao of Screen: In Search of the Distraction-Free Desktop," *Slate* (January 24, 2008), online at http://www.slate.com/articles/technology/technology/2008/01/the_tao_of_screen.html.

Jesse Grosjean explained that while working on his outliner, he came across an idea for a full-screen editor that made your computer behave like a typewriter. Blockwriter, as the imaginary program was called, disabled other programs on your computer and didn't even have delete or edit functions; like doing the crossword puzzle in ink, you could go only forward, never back. He also found Ulysses, a writing program with a full-screen mode created by German developers the Soulmen (small software companies can retain some hacker whimsy). "In the end, WriteRoom split the difference" between Ulysses and Blockwriter, he recalled. "It provided full-screen mode without tying it to a bigger 'system' such as Ulysses, but it didn't try to constrain you, as was central to the Blockwriter idea." Ulysses and other early full-screen writing tools are like Delta bluesmen to WriteRoom's Elvis. The former are often thoughtful, well executed, and have dedicated fans, but WriteRoom is the breakout.

Reviews of Zenware quoted in this chapter are Mike Gorman, "Ommwriter: Like Writing in a Zen Garden," *Geek-Tank* (September 17, 2010), online at http://www.geek-tank.com/software/ommwriter-like-writing-in-a-zen-garden/; Donald Latumahina, "Creative Thinking Cool Tool: JDarkRoom," *Life Optimizer* (February 15, 2007), online at www.lifeoptimizer.org/2007/02/15/creative-thinking-cool-tool-jdarkroom/; J. Dane Tyler, "Software Review: DarkRoom v. JDarkRoom," *Darcknyt* (December 29, 2007), online at http://darcknyt.wordpress.com/2007/12/29/software-review-darkroom-v-jdarkroom/; Richard Norden on the WriteMonkey Web site, http://writemonkey.com/; Rob Pegoraro, "That Green Again," *Washington Post* (March 20, 2008), online at http://www.washingtonpost.com/wp-dyn/content/article/2008/03/19/AR2008031903559.html.

James Anderson's work includes Catherine Weir, James Anderson, and Mervyn Jack, "On the Role of Metaphor and Language in Design of Third Party Payments in eBanking: Usability and Quality," *International Journal of*

Human-Computer Studies 64, no. 8 (2006): 770–84, and Anderson's "If Knowledge Then God: The Epistemological Theistic Arguments of Plantinga and Van Til," *Calvin Theological Journal* 40, no. 1 (2005): 49–75. Formerly a research fellow at the University of Edinburgh's Centre for Communication Interface Research, he is now a professor of theology and philosophy at the Reformed Theological Seminary in Charlotte, North Carolina.

Fred Stutzman talks about Freedom and Zenware in "Productivity in the Age of Social Media" in R. Trebor Scholz, ed., *The Digital Media Pedagogy Reader* (New York: Institute for Distributed Creativity, Comment Press, 2011), online at http://learningthroughdigitalmedia.net/productivity-in-the-age-of-social-media-freedom-and-anti-social.

Taylorism's place in the history of American technology is discussed in Thomas Parke Hughes, *American Genesis: A Century of Invention and Technological Enthusiasm, 1870–1970* (Chicago: University of Chicago Press, 1990).

George Lakoff has written extensively about framing. His *Metaphors We Live By,* coauthored with Mark Johnson (Chicago: University of Chicago Press, 1980), contains some of Lakoff's earlier thinking about framing. It also influenced the Macintosh interface design team's thinking about metaphors—Chris Espinosa recalls seeing copies of Lakoff's books on desks at Apple in the early 1980s. More recently, Lakoff has been applying his ideas to the political realm; see his books *Thinking Points* (New York: Farrar, Straus and Giroux, 2005) and *The Political Mind* (New York: Viking, 2008).

Chapter 3: *Meditate*

There is a vast literature on meditation. My own practice has been informed by Steve Hagen, *Buddhism Is Not What You Think: Finding Freedom Beyond Beliefs* (New York: HarperCollins, 2004), which is a good philosophical introduction to Buddhism; the Dalai Lama, *How to Practice: The Way to a Meaningful Life,* trans. and ed. Jeffrey Hopkins (New York: Atria Books, 2003); Katsuki Sekida, *Zen Training: Methods and Philosophy* (Boston: Shambhala, 1985), which is especially good for its examination of sitting and breathing and the role they play in good meditation; and the surprisingly deep and useful Stephan Bodian, *Meditation for Dummies* (New York: Wiley, 2006). (I *did* say I wasn't a very profound meditator.)

Joanna Cook, *Meditation in Modern Buddhism: Renunciation and Change in Thai Monastic Life* (Cambridge: Cambridge University Press, 2010), is a good introduction to contemporary Buddhism and meditation.

On the application of meditation and mindfulness, see the work of Jon Kabat-Zinn, *Full Catastrophe Living: Using the Wisdom of Your Body and Mind to Face Stress, Pain, and Illness* (New York: Random House, 1990). His "Mindfulness-Based Interventions in Context: Past, Present, and Future," *Clinical Psychology: Science and Practice* 10, no. 2 (Summer 2003): 144–56, remains a useful overview of applications of mindfulness-based stress reduction (MBSR). For a critique of MBSR, see Wakoh Shannon Hickey, "Meditation as Medicine: A Critique," *CrossCurrents* (June 2010): 168–84.

More specialized studies include William S. Blatt, "What's Special about Meditation? Contemplative Practice for American Lawyers," *Harvard Negotiation Law Review* 7 (2002): 125–41; Major G. W. Dickey, "Mindfulness-Based Cognitive Therapy as a Complementary Treatment for Combat/Operation Stress and Combat Post-Traumatic Stress Disorder" (master's thesis, Marine Corps University, 2008), online at http://www.dtic.mil/cgi-bin/GetTRDoc?AD=ADA490935&Location=U2&doc=GetTRDoc.pdf; Charlotte J. Haimer and Elizabeth R. Valentine, "The Effects of Contemplative Practice on Intrapersonal, Interpersonal, and Transpersonal Dimensions of the Self-Concept," *Journal of Transpersonal Psychology* 33, no. 1 (2001): 33–52; Keith A. Kaufman, Carol R. Glass, and Diane B. Arnkoff, "Evaluation of Mindful Sport Performance Enhancement (MSPE): A New Approach to Promote Flow in Athletes," *Journal of Clinical Sports Psychology* 4 (2009): 334–56; Ying Hwa Kee and C. K. John Wang, "Relationships Between Mindfulness, Flow Dispositions and Mental Skills Adoption: A Cluster Analytic Approach," *Psychology of Sport and Exercise* 9, no. 4 (July 2008): 393–411; Maria Lichtmann, *The Teacher's Way: Teaching and the Contemplative Life* (Mahwah, NJ: Paulist Press, 2005); Donald R. Marks, "The Buddha's Extra Scoop: Neural Correlates of Mindfulness and Clinical Sport Psychology," *Journal of Clinical Sports Psychology* 2, no. 3 (August 2008): 216–41; Ed Sarath, "Meditation in Higher Education: The Next Wave?" *Innovative Higher Education* 27, no. 4 (2003): 215–23.

Mindfulness-based legal practice argues against the usual approach to legal negotiations. Disputes are commonly seen as zero-sum games, and an adversarial approach to other parties is designed into the legal system, an attitude that discourages attorneys from looking for win-win situations, outcomes that provide some benefit to all parties. David Hoffman, "The Future of ADR: Professionalization, Spirituality, and the Internet," *Dispute Resolution Magazine* 14 (2008): 6–10; Marjorie A. Silver, "Lawyering and Its

Discontents: Reclaiming Meaning in the Practice of Law," *Touro Law Review* 19 (2004): 773–824; Arthur Zajonc, "Contemplative and Transformative Pedagogy," *Kosmos Journal* 5, no. 1 (Fall/Winter 2006): 1–3.

For an overview of neuroscientific work on meditation and consciousness, see Antoine Lutz, John D. Dunne, and Richard J. Davidson, "Meditation and the Neuroscience of Consciousness: An Introduction," in Philip David Zelazo, Morris Moscovitch, Evan Thompson, eds., *The Cambridge Handbook of Consciousness* (Cambridge: Cambridge University Press, 2007).

The studies described here are in Antoine Lutz et al., "Long-Term Meditators Self-Induce High-Amplitude Gamma Synchrony During Mental Practice," *Proceedings of the National Academy of Sciences* 101, no. 46 (November 16, 2004): 16369–73; Richard J. Davidson and Antoine Lutz, "Buddha's Brain: Neuroplasticity and Meditation," *IEEE Signal Processing Magazine* (September 2007): 171–74; Antoine Lutz et al., "Attention Regulation and Monitoring in Meditation," *Trends in Cognitive Sciences* 12, no. 4 (April 2008): 163–69.

A good introduction to studies of musicians' brains is Daniel Levitin, *This Is Your Brain on Music: The Science of a Human Obsession* (New York: Plume, 2006). For more detailed studies, see G. Schlaug et al., "In Vivo Evidence of Structural Brain Asymmetry in Musicians," *Science* 267, no. 5198 (February 3, 1995): 699–701; Stefan Elmer, Martin Meyer, and Lutz Jäncke, "Neurofunctional and Behavioral Correlates of Phonetic and Temporal Categorization in Musically Trained and Untrained Subjects," *Cerebral Cortex* 22, no. 3 (March 2012): 650–58 (doi: 10.1093/cercor/bhr142); Patrick Bermudez et al., "Neuroanatomical Correlates of Musicianship as Revealed by Cortical Thickness and Voxel-Based Morphometry," *Cerebral Cortex* 19, no. 7 (July 2009): 1583–96 (doi: 10.1093/cercor/bhn196). Mathematicians' brains are the subject of K. Aydin et al., "Increased Gray Matter Density in the Parietal Cortex of Mathematicians: A Voxel-Based Morphometry Study," *American Journal of Neuroradiology* 28 (November 2007): 1859–64. Changes in the white matter microstructure of jugglers' brains is discussed in Jan Scholz et al., "Training Induces Changes in White Matter Architecture," *Nature Neuroscience* 12, no. 11 (November 2009): 1370–71. As Scholz explains, "After six weeks of juggling training, we saw changes in the white matter of this group compared to the others who had received no training. The changes were in regions of the brain which are involved in reaching and grasping in the periphery of vision" (Scholz quoted in "Matter in Hand: Jugglers Have Rewired Brains," Phys.org (October 11, 2009), online at http://phys.org/news174490349.html#nRlv. London cabdrivers have been the subject of numerous studies, mainly conducted by Eleanor A. Maguire, a professor at University College London; see E. Maguire, Richard Frackowiak, and Christopher Frith, "Recalling Routes Around London: Activation of the Right Hippocampus in Taxi Drivers," *Journal of Neuroscience* 17, no. 18 (September 15, 1997): 7103–10; E. Maguire et al., "Navigation-Related Structural Change in the Hippocampi of Taxi Drivers," *Proceedings of the National Academy of Sciences* 97, no. 8 (April 11, 2000): 4398–403; Katherine Woollett, Hugo J. Spiers, and E. Maguire, "Talent in the Taxi: A Model System for Exploring Expertise," *Philosophical Transactions of the Royal Society, Biological Sciences* 364, no. 1522 (May 27, 2009): 1407–16.

The work of Clifford Saron and his group is summarized in the Center for Mind and Brain, *Five Year Report 2003–2008* (University of California, Davis, 2008). The telomere work is reported in Elissa Epel et al., "Can Meditation Slow Rate of Cellular Aging? Cognitive Stress, Mindfulness, and Telomeres," *Annals of the New York Academy of Sciences* 1172 (2009): 34–53.

A different but equally valuable study of the intersection of neuroscience and consciousness is Susan Blackmore, *Zen and the Art of Consciousness* (Oxford: OneWorld, 2011). Blackmore is a neuroscientist, and the book examines questions about the nature of consciousness drawing on both her scientific work and her years of meditation practice.

Material on monastic bloggers comes largely through interviews with Jonathan Coppola, Caine Das, Sister Gryphon, Choekyi Libby, Bhikkhu Samahita, Damchoe Wangmo, and Noah Yuttadhammo conducted via e-mail or Skype in the summer and fall of 2011. A study of monks and blogging, focusing in particular on Korean monastic bloggers, is Joonseong Lee, "Cultivating the Self in Cyberspace: The Use of Personal Blogs Among Buddhist Priests," *Journal of Media and Religion* 8 (2009): 97–114.

On Buddhism and globalization, see Peter Oldmeadow, "Tibetan Buddhism and Globalisation," in Carole M. Cusack and Peter Oldmeadow, eds., *The End of Religions? Religion in an Age of Globalisation,* Sydney Studies in Religion, vol. 4 (Sydney: University of Sydney, 2001), 266–79.

The life of forest monks sounds timeless, and Sri Lanka has been a center of Buddhism for over two thousand years. But the forest-monk movement, with its emphasis on isolation and purity, is fairly contemporary, supercharged by a post–World War II religious revival inspired by Sri Lanka's independence from Great Britain, in 1948, as well as the two thousand five hundredth anniversary of the Buddha's death, in 1956. Background on Sinhalese forest monks comes from Nur Yalman, "The Ascetic Buddhist Monks of Ceylon," *Ethnology* 1, no. 3 (July 1962): 315–28, and Michael Carrithers, "The Modern Ascetics of Lanka and the Pattern of Change in Buddhism," *Man* 14, no. 2 (June 1979): 294–310.

It's common practice in Buddhism to give newly ordained monks new names. These dharma names are chosen by teachers or abbots, and there are (of course) a variety of traditions for choosing them. In some cases, a teacher will choose a name that reflects a student's personality, describes the quality of his practice, or reminds him of a skill he needs to cultivate. In other traditions, names are crafted to reflect a monk's lineage or derive from his birthdate or generation. In China and Vietnam, monastic names begin with Shi or Thich, respectively.

The Axial Age is the subject of Karl Jaspers book *Origin and Goal of History* (1951; repr. London: Routledge, 2011); and more recently Karen Armstrong's *The Great Transformation: The Beginning of Our Religious Traditions* (New York: Anchor, 2007).

Scholars have argued that the Web is but the latest information technology to change our brains. The invention of writing—particularly the development of the Greek alphabet, the first that could accurately reproduce the full range of a language's sounds—profoundly altered the way humans thought. The printing press set information free five hundred years before the Internet, while the newspaper was the first near real-time medium and a critical foundation for the growth of "imagined communities." The radio, telephone, and television began to transform the world into a "global village" in the 1960s, according to Marshall McLuhan.

Humans have been through multiple information revolutions throughout history, and every generation has lamented the changes. Socrates distrusted the new medium of writing. In 1477, the Venetian humanist Hieronimo Squarciafico complained in his *Memory and Books,* "Abundance of books makes men less studious; it destroys memory and enfeebles the mind by relieving it of too much work." A hundred and fifty years ago, the telegraph was the "Victorian Internet," as described in Tom Standage's delightful *The Victorian Internet: The Remarkable Story of the Telegraph and the Nineteenth Century's On-Line Pioneers* (London: Walker, 1998); the Victorians talked about the telegraph in the same apocalyptic and prophetic tones some use to describe the Web today.

Chapter 4: *Deprogram*

On Moore's law and the history of computing, see Martin Campbell-Kelly and William Aspray, *Computer: A History of the Information Machine* (New York: Basic Books, 1996).

I'm in my twelfth turn of Moore's law. My first computer was a Macintosh Plus. When I bought it in 1988, personal computers had been around for a decade, and the Macintosh line—the first widely accessible computer with a graphical user interface and mouse—was four years old. The Plus had an 8 MHz processor, 1 MB of RAM, an 800 K floppy disk drive, and a 9-inch monochrome display. I used the Mac Plus to write my dissertation and to play more rounds of Dark Castle than I want to think about. Twenty-three years later, I bought an iPad 2 to write a book. The iPad 2 is just a little more impressive than the Plus. It has an 800 MHz dual core processor, 512 MB of RAM, and 64 GB of memory; the screen is nearly the same size as the Mac's, but it's has color and is touch-sensitive. My latest computer is at least a hundred times faster than my first and has many hundred times as much memory, and yet it cost less than my first one: the Mac Plus set me back about $2,000 in 1988 (about $3,800 in 2011 dollars), while the iPad cost around $1,000, and that's with a keyboard and other peripherals (about $525 in 1988 dollars). The Mac Plus didn't have wi-fi, and Apple didn't assume that my computer would ever be connected to the Internet. At the time, a 48.8 Kbps external modem cost a couple hundred dollars. My iPad is a hybrid: it has about 20 GB of music and movies (the equivalent of 25,000 floppy disks), but most work-related things I do on the iPad need an Internet connection. It's a terminal connected to the Cloud. The iPad is much more powerful than my Mac Plus, but more important, that power is augmented a billion times by its ability to tap into the globally distributed processing power and memory of the Web, and that power grows every day.

Much of the brain's growth happens before we are born, with a second rapid growth period in early childhood: John Dobbing and Jean Sands, "Quantitative Growth and Development of Human Brain," *Archives of Disease in Childhood* 48 (1973): 757–67.

Byron Reeves and Clifford Nass pioneered the study of our psychological responses to computers, and their book *The Media Equation: How People Treat Computers, Television, and New Media Like Real People and Places* (Cambridge: Cambridge University Press, 1996) is a good introduction to the subject. Clifford Nass's *The Man Who Lied to His Laptop: What Machines Teach Us About Human Relationships* (New York: Penguin 2010) is also an excellent, accessible work. More detailed studies are Clifford Nass, Youngme Moon, and Paul Carney, *Are People Polite to Computers? Responses to Computer-Based Interviewing Systems* (Cambridge, MA: Division of Research, Harvard Business School, 1998); Clifford Nass and Youngme Moon, "Machines and Mindlessness: Social Responses to Computers," *Journal of Social Issues* 56, no. 1 (2000): 81–103; Yasuhiro Katagiri, Clifford Nass, and Yugo Takeuchi, "Cross-Cultural Studies of the Computers Are Social Actors Paradigm: The Case of Reciprocity," in Michael Smith et al., eds., *Usability Evaluation and Interface Design: Cognitive Engineering, Intelligent Agents and Virtual Reality* (Mahwah, NJ: Lawrence Erlbaum, 2001), 1558–62.

A good overview of psychological studies with avatars is Jim Blascovich and Jeremy Bailenson, *Infinite Reality: Avatars, Eternal Life, New Worlds, and the Dawn of the Virtual Revolution* (New York: William Morrow, 2011) and Jeremy N. Bailenson and Andrew C. Beall, "Transformed Social Interaction: Exploring the Digital Plasticity of Avatars," in R. Schroeder and A. S. Axelsson, eds., *Avatars at Work and Play* (New York: Springer, 2006), 1–16. On their use in social sciences, see Jesse Fox, Dylan Arena, and Jeremy N. Bailenson, "Virtual Reality: A Survival Guide for the Social Scientist," *Journal of Media Psychology* 21, no. 3 (2009): 95–113.

The Jeremy Bailenson app includes a brief biography, PDFs of papers, directions to his laboratory at Stanford, and recent tweets. It was developed by UC-Berkeley students Shourya Basu and Jon Noreika.

AutoCAD stands for "automatic computer-aided drafting/designing"; it creates files for two-dimensional or three-dimensional objects and was originally used in architecture and precision manufacturing.

In the 1980s, military psychologists and engineers studied simulator sickness experienced by pilots training in flight simulators (which offered a full-body immersive experience), and the term *cybersickness* evolved out of this research. Human factors researchers Michael McCauley and Thomas Sharkey coined the term in their 1992 paper "Cybersickness: Perception of Self-Motion in Virtual Environments," *Presence* 1, no. 3 (1992): 311–18. As McCauley later recalled, "The term 'cyber' was going viral at the time and we were doing research on 'simulator sickness' and 'VE or VR sickness.' It seemed like an obvious step to call it 'cybersickness'" (McCauley, e-mail to the author, July 2, 2012).

Specific studies discussed here are Jeremy Bailenson et al., "Transformed Social Interaction: Decoupling Representation from Behavior and Form in Collaborative Virtual Environments," *Presence* 13, no. 4 (August 2004): 428–41; Nick Yee and Jeremy Bailenson, "The Proteus Effect: The Effect of Transformed Self-Representation on Behavior," *Human Communication Research* 33 (2007): 271–90; Jeremy Bailenson et al., "Facial Similarity between Voters and Candidates Causes Influence," *Public Opinion Quarterly* 72 (2008): 935–61; Sun Joo Ahn and Jeremy Bailenson, "Self-Endorsing Versus Other-Endorsing in Virtual Environments: The Effect on Brand Attitude and Purchase Intention," *Journal of Advertising* 40, no. 2 (Summer 2011): 93–106.

Daryl Bem's observation comes from Bem, "Self-Perception Theory," in Leonard Berkowitz, ed., *Advances in Experimental Social Psychology,* vol. 6 (New York: Academic Press, 1972), 2–57.

There's a substantial literature on future selves and planning; I review it in my "Futures 2.0: Rethinking the Discipline," *Foresight: The Journal of Futures Studies, Strategic Thinking and Policy* 12, no. 1 (Spring 2010): 5–20. In philosophy, the work of Derek Parfit on future selves (which Hal Hershfield tests) looms especially large: see Derek Parfit, *Reasons and Persons* (Oxford: Oxford University Press, 1984).

Jesse Fox's work is discussed in two articles coauthored with Bailenson: "Virtual Virgins and Vamps: The Effects of Exposure to Female Characters' Sexualized Appearance and Gaze in an Immersive Virtual Environment," *Sex Roles* 61 (2009): 147–57, and "Virtual Self-Modeling: The Effects of Vicarious Reinforcement and Identification on Exercise Behaviors," *Media Psychology* 12 (2009): 1–25.

Hal Hershfield's work is published in Hershfield et al., "Neural Evidence for Self-Continuity in Temporal Discounting," *Social Cognitive and Affective Neuroscience* 4, no. 1 (2009): 85–92; "Don't Stop Thinking about Tomorrow: Individual Differences in Future Self-Continuity Account for Saving," *Judgment and Decision Making* 4, no. 4 (2009): 280–86; and "Increasing Saving Behavior Through Age-Progressed Renderings of the Future Self," *Journal of Marketing Research* 48 (November 2011): S23–37. The questions experimental subjects were asked were simple ones like "What is your name?" "Where are you from?" and "What is your passion in life?"

On failure, see Charles Perrow, *Normal Accidents: Living with High Risk* (New York: Basic, 1984). See also Mica Endsley, "Automation and Situation

Awareness," in R. Parasuraman and M. Mouloua, eds., *Automation and Human Performance: Theory and Applications* (Mahwah, NJ: Lawrence Erlbaum, 1996), 163–81, which describes how automation can dampen users' understanding of the world around them and the technologies they're managing.

My thinking about failure and technical error in computing is influenced by the work of my fellow Microsoft Research Cambridge alum Helena Mentis; see her "User Recalled Occurrences of Usability Errors: Implications on the User Experience," *CHI '03: New Horizons* (2003): 736–37, and her "Occurrence of Frustration in Human-Computer Interaction: The Affect of Interrupting Cognitive Flow" (master's thesis, Cornell University, 2004).

Jaron Lanier's critique of crowdsourcing is from his terrific polemic, *You Are Not a Gadget: A Manifesto* (New York: Random House, 2010).

Ray Kurzweil's *The Singularity Is Near* (New York: Viking, 2005) is both very technical and surprisingly accessible. As should be clear, I disagree with most of its premise, and I find its breezy "There's no reason not to believe that the complicated philosophical problem I've just described won't have a brute-force computational solution in a few years" arguments unconvincing, but it's still, strangely, worth reading.

Gordon Bell lifelogging manifesto *Your Life, Uploaded* (New York: Plume, 2010), coauthored with Jim Gemmell, should be balanced with reflections from the HCI literature on how human and computer memory differ. In particular, the work of Abigail J. Sellen, another Microsoft Research colleague, is very much worth reading; see her article with Steve Whittaker, "Beyond Total Capture: A Constructive Critique of Lifelogging," *Communications of the ACM* 53, no. 5 (May 2010): 70–77, and Vaiva Kalnikaite et al., "Now Let Me See Where I Was: Understanding How Lifelogs Mediate Memory," *CHI '10: Remember and Reflect* (Atlanta, GA: April 10–15, 2010): 2045–54. A broad, thoughtful critique of assumptions that we can automate or replace human skills in the near future is Richard Harper et al., eds., *Being Human: Human Computer Interaction in the Year 2020* (Cambridge: Microsoft Research Ltd., 2008).

Digital memory follows a curve similar to Moore's law. When I was in college, I visited a researcher studying social networks in science who used most of his government research grant to buy a twenty-megabyte hard drive. Today, memory cards a hundred times bigger are bundled with digital cameras and given away.

Viktor Mayer-Schönberger's *Delete: The Virtue of Forgetting in the Digital Age* (Princeton, NJ: Princeton University Press, 2009) is an eloquent study of the differences between digital and human memory. One good example of the social nature of memory is arrest records. It's getting harder to expunge arrest records and keep crimes you've already paid the price for from haunting you: there are often multiple copies of records, and those are not always under central control. To make things more complicated, there are now companies that specialize in putting criminal records online as well as companies that offer to scrub those records (sometimes, it seems, they're run by the same people).

Morgan Ames's work on One Laptop Per Child is documented in Mark Warschauer and Morgan Ames, "Can One Laptop Per Child Save the World's Poor?" *Journal of International Affairs* 64, no. 1 (Fall/Winter 2010): 33–51; and Mark Warschauer et al., "One Laptop per Child Birmingham: Case Study of a Radical Experiment," *International Journal of Learning and Media* 3, no. 2 (Spring 2011): 61–76. On hackers, see Steven Levy's 1984 *Hackers: Heroes of the Computer Revolution* (repr. Sebastopol, CA: O'Reilly Media, 2010). It's one of those books that first describe a culture and then influence it (rather like how the Godfather movies changed the way wiseguys behave). Pekka Himanen's *The Hacker Ethic* (New York: Random House, 2001) is also quite good. Among Claude Steele's numerous articles, a good introduction is Steele and Joshua Aronson, "Stereotype Threat and the Intellectual Test Performance of African Americans," *Journal of Personality and Social Psychology* 69, no. 5 (1995): 797–811. Also valuable are Steele's "A Threat in the Air: How Stereotypes Shape Intellectual Identity and Performance," *American Psychologist* 52, no. 6 (June 1997): 613–29, and his *Whistling Vivaldi: How Stereotypes Affect Us and What We Can Do* (New York: W. W. Norton, 2010). The impact of fixed versus growth mindsets is examined in Carol Dweck, *Mindset: The New Psychology of Success* (New York: Random House, 2006).

Chapter 5: *Experiment*

According to a 2008 AOL survey, 59 percent of people report checking e-mail in the bathroom. (AOL Mail fourth annual e-mail addiction survey, 2008; results online at http://cdn.webmail.aol.com/survey/aol/en-us/index.htm.) Among BlackBerry users, 91 percent admit checking e-mail in the bathroom; see Kevin Michaluk, Martin Trautschold, and Gary Mazo, *CrackBerry: True Tales of BlackBerry Use and Abuse* (New York: Apress, 2010), 16–17.

Self-experimentation is described in Seth Roberts, "Self-Experimentation as a Source of New Ideas: Ten Examples about Sleep, Mood, Health, and Weight," *Behavioral and Brain Sciences* 27 (2004): 227–88.

Tinkering is more than customization or reading the manual. It's a pragmatic, improvisational approach to engaging with and changing technologies, and it emphasizes flexibility, rapid learning, and bricolage. It can have a playful, engaging aspect that some describe as Zen-like. Tinkering is also quite social: tinkerers share their ideas with one another, trade design tips, and show off their work. In the United States, tinkering has been elevated to a form of self-education and self-improvement; it's a playful way to learn new skills and more fully understand one's built environment. Mark Frauenfelder's *Made by Hand: Searching for Meaning in a Throwaway World* (New York: Portfolio, 2010) is a great introduction to tinkering, coming as it does from the editor of *Make* magazine and cofounder of the Maker Faire. For more academic perspectives, see Anne Balsamo, *Designing Culture: The Technological Imagination at Work* (Durham, NC: Duke University Press, 2011), especially chapter 4, and Anya Kamenetz, *DIY U: Edupunks, Edupreneurs, and the Coming Transformation of Higher Education* (White River Junction, VT: Chelsea Green, 2010).

Rupert Brooke described Grantchester in his poem "The Old Vicarage, Grantchester," written in 1912.

Thomas Merton's contemplative photography is the subject of Philip Richter's article "Late Developer: Thomas Merton's Discovery of Photography as a Medium for His Contemplative Vision," *Spiritus: A Journal of Christian Spirituality* 6, no. 2 (Fall 2006): 195–212.

An interesting study of mindfulness and gaming is Jayne Gackenbach and Johnathan Bown, "Mindfulness and Video Game Play: A Preliminary Inquiry," *Mindfulness* 2, no. 2 (June 2011): 114–22.

My understanding of affordances derives from Abigail Sellen and Richard Harper's book *The Myth of the Paperless Office* (Cambridge, MA: MIT Press, 2001), which brilliantly shows how features of printed media that we didn't think were important turn out to be the foundations for all kinds of reading and working practices.

Vannevar Bush describes the memex in "As We May Think," *Atlantic Monthly* (July 1945), available online at http://www.theatlantic.com/magazine/archive/1969/12/as-we-may-think/3881/. While the memex was never built, it is now considered one of the earliest and most inspiring descriptions of hypertext; the impact of Bush's essay is described in James Nyce and Paul Kahn, eds., *From Memex to Hypertext: Vannevar Bush and the Mind's Machine* (San Diego: Academic Press, 1991).

The intensive readers I interviewed are increasingly in the minority when it comes to avoiding children's e-books, as the market for both children's titles that can be read on the Kindle or iPad and dedicated devices by companies like VTech is growing rapidly. However, there is now evidence that children learn to read more quickly and effectively with printed books; see Cynthia Chiong et al., *Print Books vs. E-Books: Comparing Parent-Child Co-Reading on Print, Basic, and Enhanced E-Book Platforms* (New York: Joan Ganz Cooney Center at Sesame Workshop, 2012), available online at http://www.joanganzcooneycenter.org.

The phrase *ironies of automation* comes from the title of a classic article by Lisanne Bainbridge, published in *Automatica* 19, no. 6 (November 1983): 775–79, which argued that "the more advanced a control system is, so the more crucial may be the contribution of the human operator." On the history of household technologies, Ruth Schwartz Cowan's *More Work for Mother: The Ironies of Household Technology from the Open Hearth to the Microwave* (New York: Basic Books, 1985) is still a must-read.

The Jevons paradox is first described in William Stanley Jevons, *The Coal Question: An Inquiry Concerning the Progress of the Nation, and the Probable Exhaustion of Our Coal-Mines* (London: Macmillan and Co., 1865), especially chapter 7.

The failure of antilock brakes to reduce accidents and the theory that drivers were engaging in "risk compensation"—that is, driving more aggressively because they perceived themselves to be safer—was first noted in M. Aschenbrenner and B. Biehl's "Improved Safety Through Improved Technical Measures? Empirical Studies Regarding Risk Compensation Processes in Relation to Anti-Lock Brake Systems," in R. M. Trimpop and G. J. S. Wilde, eds., *Changes in Accident Prevention: The Issue of Risk Compensation* (Groningen, the Netherlands: Styx Publications, 1994), 81–89. Edward Tenner's *Why Things Bite Back: Technology and the Revenge of Unintended Consequences* (New York: Vintage, 1997) also describes this and many other examples of technologies bringing unintended consequences. See also Alex Soojung-Kim Pang, "A Banquet of Consequences: Living in the 'Nobody-Could-Have-Predicted' Era," *World Future Review* 3, no. 2 (Summer 2011): 5–10.

Marguerite Manteau-Rao's blog Mind Deep (http://minddeep.blogspot.com/) is well written, confessional, and smart.

Elizabeth Drescher's *Tweet If You Heart Jesus: Practicing Church in the Digital Reformation* (Harrisburg, PA: Morehouse, 2011) is one of a large number of books about social media and the modern church; Jesse Rice's *The Church of Facebook* (Colorado Springs, CO: David C. Cook, 2009) is also good. Soren Gordhamer's *Wisdom 2.0: Ancient Secrets for the Creative and Constantly Connected* (New York: HarperOne, 2009) and Lori Deschene's "Ten Mindful Ways to Use Social Media: Right Tweeting Advice from @TinyBuddha," *Tricycle* (Spring 2011), online at http://www.tricycle.com/feature/ten-mindful-ways-use-social-media, both apply Buddhist philosophy to dealing with technology.

That suggestion to live first and tweet later calls to mind literary historian Walter Ong's observation that stories became more complicated with the rise of print culture. Oral tales tend to be linear descriptions of events; the structure is some variant on "this happened, then that happened, then another thing happened." It's only with writing that you get sophisticated narrative forms and the awareness that the interpretation of events can change over time, making storytelling more complex. The real-time emphasis of social media makes it seem more conversational, but as it re-creates some elements of oral culture, it may also encourage us to be a bit less reflective. See Ong, *Orality and Literacy.*

The Renaissance commonplace book is discussed in Ann Blair's "Humanist Methods in Natural Philosophy: The Commonplace Book," *Journal of the History of Ideas* 53, no. 4 (October 1992): 541–51, and Ann Moss's *Printed Commonplace-Books and the Structuring of Renaissance Thought* (Oxford: Clarendon Press, 1996).

The discussion of architecture and drawing cites Witold Rybczynski, "Think Before You Build: Have Computers Made Architects Less Disciplined?" *Slate* (March 30, 2011), online at http://www.slate.com/articles/arts/architecture/2011/03/think_before_you_build.html; an interview with Renzo Piano in *Architectural Record* (2011), online at http://archrecord.construction.com/people/interviews/archives/0110piano.asp; interviews with William Huchting, David Brownlee, and Chris Luebkeman. See also James Wines, "Drawing and Architecture," *Blueprint* (September 30, 2009), online at http://www.blueprintmagazine.co.uk/index.php/architecture/james-wines-drawing-and-architecture/.

Complaints about the detrimental effects of CAD on education and thinking are not confined to Penn. Professor Alan Balfour contends that before computers, students had to draw on history, sculpture, design books, and one another; the digital world, by contrast, "is an internalized, constrained, and virtual experience in which the creative relationship to the tools and information held within the machine seem to be more stimulating and to hold more promise than the experience of place, or the lessons of history": Balfour, "Architecture and Electronic Media," *Journal of Architectural Education* 54, no. 4 (May 2001): 268–71. Likewise, Syracuse University professor Robert Svetz argues that "the digital productivity mindset exacts too high a design price when it displaces better equipped modes of manually graphic thinking, and learning": Svetz, "Drawing/Thinking: Confronting an Electronic Age," *Journal of Architectural Education* 63, no. 1 (October 2009): 155–57.

Knuth's good-bye to e-mail is, of course, online; see http://www-cs-faculty.stanford.edu/~uno/email.html.

Chapter 6: *Refocus*

The level problem with Do Nothing for Two Minutes was pointed out to me by editor Susana Darwin.

James Watson's *The Double Helix: A Personal Account of the Discovery of the Structure of DNA* was first published in 1968; the 2001 Touchstone edition, with a preface by Sylvia Nasar, is the one I carried to Cambridge.

In a 2008 *Nature* survey, 20 percent of readers said they had "used drugs for nonmedical reasons to stimulate their focus, concentration or memory." Brendan Maher, "Poll Results: Look Who's Doping," *Nature* 452 (April 10, 2008): 674–75. The poll was inspired by Barbara Sahakian and Sharon Morein-Zamir, "Professor's Little Helper," *Nature* 450 (December 20, 2007): 1157–59. The use of pharmaceuticals to boost brainpower is not confined to faculty members or graduate students; polls indicate that cognitive enhancers are becoming more popular among American undergraduates as well. See Beth Azar, "Better Studying Through Chemistry?" *APA Monitor* 39, no. 8 (September 2008): 42.

In recent years, Darwin has attracted remarkable biographers who've written excellent books; Janet Browne's two-volume account, *Charles Darwin: Voyaging* (Princeton, NJ: Princeton University Press, 1996) and *Charles*

Darwin: The Power of Place (Princeton, NJ: Princeton University Press, 2003), and Adrian Desmond and James Moore's *Darwin: The Life of a Tormented Evolutionist* (New York: W. W. Norton, 1994) are both beautifully written pieces of scholarship and are the foundations for my understanding of Darwin.

The voyage of the HMS *Beagle* is rightly remembered as a turning point in both Darwin's life and the history of science. For five years, Darwin conducted geological and biological fieldwork in regions little studied by Western scientists. While the *Beagle* conducted its surveys, Darwin would go ashore, collect specimens, and make observations. What he saw provided fodder for decades of thinking and theorizing. He witnessed an earthquake in Chile that provided evidence that natural forces were constantly acting on the world, steadily and incrementally thrusting up mountain chains and subsiding ocean volcanoes; divine causes like the biblical Flood could be disregarded. He surmised that coral atolls in the Pacific were formed on the rims of submerged volcanoes. Most famously, in the Galapagos Islands, he observed the differentiation of species, which began a line of thought that concluded years later with his formulation of the theory of evolution by natural selection.

Darwin was able to settle into (and support) his life as a country gentleman in part because he and Emma came from money; one of their grandfathers (Charles and Emma were cousins) had founded the Wedgwood pottery business, and Charles's father, Robert, had been a shrewd investor in real estate and business. Life at Down House was supported by inheritances, investments, and income from the farm, and this was sufficient for a respectable if not lavish lifestyle. Down House and the Sandwalk are discussed in Arthur Keith's "Side-Lights on Darwin's Home—Down House," *Annals of the Royal College of Surgeons* 12, no. 3 (March 1953): 197–207, and in Gene Kritsky's "Down House: A Biologist's Perspective," *Bios* 54, no. 1 (March 1983): 6–9. The regularity of Darwin's life ("My life goes on like Clockwork, and I am fixed on the spot where I shall end it") is referred to in correspondence from Darwin to Robert FitzRoy, October 1, 1846. The Darwin Correspondence Project is available online at http://www.darwinproject.ac.uk.

You can get a firsthand sense of Darwin from the Darwin Correspondence Project, which has put online a number of his letters as well as related contemporary publications. Wood engravings of the Sandwalk and Down House are published in Rev. O. J. Vignoles's "The Home of a Naturalist,"

Good Words 34 (1893): 95–101, available online at http://darwin-online.org .uk/content/frameset?viewtype=side&itemID=A483&pageseq=1.

On the history of walking, see Rebecca Solnit's *Wanderlust: A History of Walking* (London: Verso, 2001). The cognitive benefits of walking come from walking's physiological effects (it enhances cognitive ability by strengthening the heart and improving blood flow, thus delivering more energy to the brain) and the psychological benefits of interacting with natural environments; see Marc G. Berman, John Jonides, and Stephen Kaplan, "The Cognitive Benefits of Interacting with Nature," *Psychological Science* 19, no. 12 (2008): 1207–12. On the therapeutic benefits of walking after brain injury, see Andreas R. Luft et al., "Treadmill Exercise Activates Subcortical Neural Networks and Improves Walking After Stroke: A Randomized Controlled Trial," *Stroke* 39, no. 12 (December 2008): 3341–50.

Stephen Kaplan's work on restorative natural environments is presented in Stephen Kaplan, Lisa V. Bardwell, and Deborah B. Slakter, "The Museum as a Restorative Environment," *Environment and Behavior* 25 (1993): 725–42; Stephen Kaplan and J. Talbot, "Psychological Benefits of a Wilderness Experience," in I. Altman and J. F. Wohlwill, eds., *Behavior and the Natural Environment* (New York: Plenum, 1993), 163–203; Stephen Kaplan, "The Restorative Benefits of Nature: Toward an Integrative Framework," *Journal of Environmental Psychology* 16 (1995): 169–82; and Stephen Kaplan, "Meditation, Restoration, and the Management of Mental Fatigue," *Environment and Behavior* 33 (2001): 480–506. The quotation is from Kaplan, "The Urban Forest as a Source of Psychological Well-Being," in Gordon Bradley, ed., *Urban Forest Landscapes: Integrating Multidisciplinary Perspectives,* (Seattle: University of Washington Press, 1995), 102.

Rebecca Krinke, ed., *Contemporary Landscapes of Contemplation* (London: Routledge, 2005), and Bianca C. Soares Moura, "Contemplation-Scapes: An Enquiry into the Strategies, Typologies, and Design Concepts of Contemplative Landscapes" (master's thesis, Edinburgh College of Art, 2009), apply Kaplan's work to the built environment and design.

The therapeutic benefits of gardens is the subject of an entire line of academic inquiry, represented by the *Journal of Therapeutic Horticulture,* published by the American Horticultural Therapy Association. In the United Kingdom, the field of social and therapeutic horticulture has developed in recent years; for an introduction, see Joe Sempik, Jo Aldridge, Saul Becker, *Health, Well-Being, and Social Inclusion: Therapeutic Horticulture in the UK*

(Bristol, UK: Policy Press, 2005). An accessible overview of the broader literature on the restorative value of nature is Eric Jaffe's "This Side of Paradise: Discovering Why the Human Mind Needs Nature," *Association for Psychological Science Observer* (May/June 2010), online at http://www.psychologicalscience.org/observer/getArticle.cfm?id=2679.

Two essays that describe restorative activity as distraction are Hanif Kureishi's "The Art of Distraction," *New York Times* (February 19, 2012), and James Surowiecki's "In Praise of Distraction," *New Yorker* (April 11, 2011).

Chapter 7: *Rest*

Most of this chapter is based on interviews. However, the digital Sabbath has attracted a few, fortunately good, writers. William Powers's *Hamlet's BlackBerry: Building a Good Life in the Digital Age* (New York: HarperCollins, 2010) closes with a chapter on them. Susan Maushart's *The Winter of Our Disconnect* (London: Profile Books, 2011) is a funny, wry account of her months unplugging her family.

In April 2012, Deion Sanders (a Football Hall of Fame cornerback and former professional baseball player) live-tweeted a domestic disturbance during which he claimed his wife, Pilar (now his ex-wife), "and a friend jump[ed] me in my room." Chuck Schilken, "Deion Sanders Tweets Wife Assaulted Him in Front of Their Kids," *Los Angeles Times* (April 24, 2012), online at http://articles.latimes.com/2012/apr/24/news/chi-deion-120424. NASCAR driver Brad Keselowski went one better, live-tweeting a car crash *during* the 2012 Daytona 500; see Bill Speros, "It's a NASCAR Social Media Meet-Up!" *ESPN* (February 28, 2012), online at http://espn.go.com/espn/page2/story/_/id/7626813/brad-keselowski-live-tweet-turns-daytona-500-social-media-meet-up. Fox News contributor Grant Cardone and CNN correspondent Ali Velshi live-tweeted an emergency landing when they were both passengers on Delta Flight 1063: "Birds' Run-In with Engine Caught on Twittersphere," *89.3 KPCC* (April 19, 2012), online at http://storify.com/kpcc/birds-run-in-with-engine-caught-on-twittersphere.

Of course, in-flight wi-fi is coming to a seat near you. The first services appeared on American flights in 2009, and a number of major international airlines announced that they would open wi-fi service on long-haul flights in 2013.

The presence of advertising executives in the digital Sabbath movement—like the founders of Offlining, Mark DiMassimo (founder and CEO of DIGO) and Eric Yaverbaum (president of Jericho Communication)—may seem odd; DiMassimo and Yaverbaum confess that they "devoted much of the last couple of decades" before starting Offlining "to convincing you to log on, click here, call now, surf, and search." But recall the production of Zenware by advertising agencies; they have problems with distraction just like the rest of us.

The first media reference to digital Sabbaths appears to be Don Lattin, "In Praise of a Day Unplugged," *SF Gate* (April 29, 2001), online at http://www.sfgate.com/living/article/In-praise-of-a-day-unplugged-2926770.php, which discusses digital Sabbaths (or data Sabbaths) in Silicon Valley.

No one has yet measured the impact of digital Sabbaths on participants' heart rates, blood pressure, or other physiological indicators of health. Demographers have long observed a correlation between health and religious observance, so such a study could be useful. Among American Jews, there is a gentle correlation between health and religious observance in self-reported studies, though it's difficult to know if the improvements come from following dietary laws, Sabbath observation, or something else; see Isaac Eberstein and Kathleen Heyman, "Jewish Identity and Self-Reported Health," in Christopher G. Ellison and Robert A. Hummer, eds., *Religion, Families, and Health: Population-Based Research in the United States* (New Brunswick, NJ: Rutgers University Press, 2010), 349–67.

David Levy's writings on technology and contemplation are worth tracking down. See in particular "To Grow in Wisdom: Vannevar Bush, Information Overload, and the Life of Leisure," *Proceedings of the 5th ACM/IEEE-CS Joint Conference on Digital Libraries* (New York: ACM, 2005): 281–86; "Information, Silence, and Sanctuary," *Ethics and Information Technology* 9 (2007): 233–36; and "No Time to Think: Reflections on Information Technology and Contemplative Scholarship," *Ethics and Information Technology* 9 (2007): 237–49. Levy now leads a research project on what he calls "contemplative multitasking": see Levy et al., "Initial Results from a Study of the Effects of Meditation on Multitasking Performance," *Extended Abstracts of the ACM Conference on Human Factors in Computing Systems (CHI '11)* (Vancouver, BC: May 7–12, 2011): 2011–16; and Levy et al., "The Effects of Mindfulness Meditation Training on Multitasking in a High-stress Information Environment," *Proceedings of Graphics Interface (GI '12)* (Toronto, Ontario: May 28–30, 2012): 45–52.

Tammy Strobel describes her efforts at simplification in *You Can Buy Happiness (and It's Cheap): How One Woman Radically Simplified Her Life and How You Can Too* (Novato, CA: New World Library, 2012). Christine Rosen has an elegant short essay on multitasking; see "The Myth of Multitasking," *New Atlantis,* no. 20 (Spring 2008): 105–10. Gwen Bell's e-book *Digital Warriorship* (2011) described her digital sabbaticals.

The idea that the Amish reject modern technology is a popular one, but in reality it's no more true for the Amish than it is for Buddhist monks. Rather, Jameson Wetmore explains, they collectively "choose technologies that they believe will best promote the values they hold most dear—values like humility, equality, and simplicity" and "that are different from those used by the outside world": Wetmore, "Amish Technology: Reinforcing Values and Building Community," *IEEE Technology and Society Magazine* (Summer 2007): 10–21, quote on page 21. More detailed studies of Amish approaches to technology have been written by Donald Kraybill in *The Riddle of Amish Culture* (Baltimore: Johns Hopkins University Press, 2001) and Kraybill and Steven Nolt, *Amish Enterprise: From Plows to Profits* (Baltimore: Johns Hopkins University Press, 2004), esp. 106–24. *The Amish Struggle with Modernity,* a collection of essays edited by Kraybill and Marc Alan Olshan (Hanover, NH: University Press of New England, 1995), also contains some excellent material. Diane Zimmerman Umble has produced a number of studies of Amish and Mennonite attitudes to technology and media. Her essay "The Amish and the Telephone: Resistance and Reconstruction," in Roger Silverstone and Eric Hirsch, eds., *Consuming Technologies: Media and Information in Domestic Spaces* (London: Routledge, 1992), 183–94, summarizes elements of her later book *Holding the Line: The Telephone in Old Order Mennonite and Amish Life* (Baltimore: Johns Hopkins University Press, 1996). Finally, for a popular treatment of the subject, see Howard Rheingold, "Look Who's Talking," *Wired* 7, no. 1 (January 1999), online at http://www.wired.com/wired/archive/7.01/amish.html.

Thomas Merton mentions beer in his essay "Contemplation in a World of Action," reprinted in Lawrence Cunningham, ed., *Thomas Merton, Spiritual Master: The Essential Writings* (Mahwah, NJ: Paulist Press, 1992), 377. Merton may have been echoing Benjamin Franklin's famous declaration that "beer is proof that God loves us and wants us to be happy," a charming but, tragically, apocryphal quip. As Charles W. Bamforth notes in *Beer Is Proof God Loves Us: Reaching for the Soul of Beer and Brewing* (Upper Saddle River, NJ: Pearson Education, 2011), Franklin did, tragically, say something like this about wine.

While the phrase *paper tweeting* is one Bell came up with herself, the term *paper tweet* appears to have been first used by Jen Bilik, founder of Knock Knock, which designed and sold a Paper Tweets notepad between 2010 and 2012. Employing the "wireless miracle of pen and paper," it let writers "master the art of brevity" and taught "modern-day etiquette for the masses"; Paper Tweets was the latest in a series of paper products that played with digital media. Others included a Paper GPS (for writing directions), Paper E-mail, and Paper Emoticons. As Bilik explains, "A lot of humor comes from the embracing of opposites or presentation of things in an unexpected format. The utility of an e-mail or tweet is in its digital nature, so the folly of doing it in paper is inherently funny"; interview with Jen Bilik, July 10, 2012; product description from the Knock Knock online catalog, http://www.knockknockstuff.com/catalog/categories/pads/nifty-notes/paper-tweet-nifty-note/. Bilik and Knock Knock are profiled in Liz Welch, "The Way I Work: Jen Bilik of Knock Knock," *Inc.* (October 2011), online at http://www.inc.com/magazine/201110/the-way-i-work-jen-bilik-of-knock-knock.html.

The Sabbath Manifesto (http://www.sabbathmanifesto.org/) consists of ten principles: avoid technology; connect with loved ones; nurture your health; get outside; avoid commerce; light candles; drink wine; eat bread; find silence; and give back.

One reason digital Sabbatarians are "mildly observant" is that the more religious techies have less need to invent new rituals or rationales. For Christians, it's not difficult to fold going offline into Sunday routines that already emphasize focusing on family and community, while Orthodox Jews have plenty of elaborate, formal prohibitions against using nonessential electronics and electrical appliances. For them, the Sabbath doesn't need to be reinvented. It just needs to be observed. Indeed, the Sabbath Manifesto and digital Sabbath movement have been the subject of criticism among more Orthodox commentators; see Joseph Aaron, "People of the Twitter," *Chicago Jewish News* (April 29, 2011), online at http://www.chicagojewishnews.com/story.htm?sid=2&id=254535. However, anecdotal evidence suggests that even in Orthodox households, digital distraction is a problem; see Steve Lipman, "For Many Orthodox Teens, 'Half Shabbos' Is a Way of Life," *Jewish Week* (June 22, 2011), online at http://www.thejewishweek.com/news/national-news/many-orthodox-teens-half-shabbos-way-life.

Morley Feinstein appears in season 5, episode 41 of *Curb Your Enthusiasm,* "The Larry David Sandwich," first aired on September 25, 2005.

My lack of faith can be blamed on my parents, whose complete inattention to faith was a reaction against their own strict upbringings. When I was a child living in rural Virginia, Sunday wasn't a spiritual day; it meant dealing with an illogical patchwork of blue laws, facing previously ignored homework, and playing the occasional round of a game my grandmother invented, called Why Isn't That Boy in Sunday School? My wife and her family describe themselves as culturally Protestant; they can trace their lineage back to the Revolution, and their Thanksgiving is something out of a Norman Rockwell painting (several families gathered around a huge table laden with food, minus the grace). I feel at religious services a bit like a deaf person must feel at the ballet: He can appreciate the dancers' athleticism and devotion to their art, admire the sets, and understand the rhythm of the story, but there's a part of the experience that's always inaccessible to him. In Cambridge, we attended Evensong in the magnificent King's College Chapel, and I loved the beautiful proceedings and the majesty of the space. The music and pace of the service were restorative; the Anglican idea of the service as a "cool and ancient order" was wonderful; and if there's any place that can rightfully be called majestic, it's the vast English Gothic chapel endowed by Henry VIII as a testament to the piety and power of the Tudor line (which had recently been enriched by wealth stolen from the Catholic Church). But even there, I never felt the presence of the Divine, the touch that the devout seem to experience and treasure. Nor, to be honest, do I think I ever have.

In 2005, Abraham Heschel's *The Sabbath: Its Meaning for Modern Man* was reprinted by Farrar, Straus and Giroux in a lovely new edition with an illuminating preface by his daughter Susannah. The Sabbath has been the subject of several other great works recently, most notably Judith Shulevitz's *The Sabbath World: Glimpses of a Different Order of Time* (New York: Random House, 2010).

Wayne Hope's "Global Capitalism and the Critique of Real Time," *Time and Society* 15, no. 2–3 (2006): 275–302, offers a left-wing critique of the idea of real time. My thinking about industrial and astronomical time is inspired by Simon Schaffer's "Astronomers Mark Time: Discipline and the Personal Equation," *Science in Context* 2 (1988): 115–45.

Of course, real time has its benefits as well as its costs. For example, in his hometown of New York City, Anthony Townsend reports that taxi drivers use mobile phones to share information about where groups of people are congregating, what streets are congested, and what routes are free. Anyone who's used real-time travel or weather updates or converged with friends after a flurry of calls and texts knows the benefits of being able to share information in real time. Sometimes it even produces delightful juxtapositions. My Facebook wall once threw out a lovely cross-section of my friend's lives: "A power trio of 6 graders just rocked Blue Oyster Cult's 'Godzilla' "; "Ah, Dublin, Dublin, Dublin"; "Hoboken Italian Festival in Sinatra Park. This is the celestial convergence of Jerseyana"; "Brenda McMorrow is rocking my Bhakti heart on the Hanuman stage." In order to find this little convergence several months after I'd first noticed it, though, I had to download a copy of my Facebook wall and search it. At the time, Facebook didn't offer the ability to search one's own wall. As far as the system was concerned, what was happening right at that present moment was all that mattered.

Daniel Sieberg's *The Digital Diet: The 4-Step Plan to Break Your Tech Addiction and Regain Balance in Your Life* (New York: Three Rivers Press, 2011) offers a twenty-eight-day program to cure your addiction to technology. It begins with a seven-day detox, recommends storing smartphones and other distracting technologies in the refrigerator, and even encourages you to measure your "virtual weight index" (the electronic equivalent of the body mass index) by adding up the number of electronic devices you own, social-media and e-mail accounts you check, virtual games you play, and blogs you write. In this system, the more distracting an object is, the more it "weighs": a smartphone weighs three times as much as a desktop computer, and an account on World of Warcraft weighs seven times as much a digital camera.

Clay Johnson's *The Information Diet: A Case for Conscious Consumption* (San Francisco: O'Reilly Media, 2012) is sort of a cross between Michael Pollan and David Broder. Johnson, a digital guru on Howard Dean's 2004 presidential campaign, argues that partisan news and blogs are the equivalent of high-calorie, low-fiber junk food and are responsible for the decline of the American political climate and the rise in voters who are simultaneously better connected and less informed.

Incredibly, there's no comparative study of fasting across different religious traditions. Until such a book appears, good information can be found in Kees Wagtendonk's *Fasting in the Koran* (Leiden, Netherlands: E. J. Brill, 1968) and the works of sociologist Joseph B. Tamney, particularly his "Fasting and Modernization," *Journal for the Scientific Study of Religion* 19, no. 2 (June 1980): 129–37, and "Fasting and Dieting: A Research Note," *Review of Religious Research* 27, no. 3 (March 1986): 255–62. The health impacts of religious fasting are described in A. M. Johnstone, "Fasting: The Ultimate Diet?" *Obesity Reviews* 8, no. 3 (May 2007): 211–22, and John F. Trepanowski and Richard J. Bloomer, "The Impact of Religious Fasting on Human Health," *Nutrition Journal* 9 (2010): 57–65.

This idea of the Sabbath rest as active is echoed by authors both inside and outside Judaism. Anne Dilenschneider, the Methodist pastor, quotes the words of David Steindl-Rast, a Viennese-born experimental psychologist and Benedictine monk: "The cure for exhaustion is not *rest*," he says. "The cure is whole-heartedness." In the original techno-Sabbath course, Dilenschneider and Bauer recounted a story from the *Likrat Shabbat* about a famous pianist who told an admirer that it was not in the notes that his art resided, but in the pauses between them. "In great living, as in great music," a rabbinical commentary says, "the art may be in the pauses." The Sabbath pause offers an opportunity to cultivate "the art of living."

Chapter 8: *Eight Steps to Contemplative Computing*

Of course, the list in this chapter mirrors the Noble Eightfold Path in Buddhism, which consists of right view, right aspiration, right speech, right action, right livelihood, right effort, right mindfulness, and right concentration.

Go (known as *weiqi* in China, and *baduk* in Korea) was invented in the sixth century BCE in China, making it one of the oldest board games still played today. Players use black and white stones to capture as much of a 19-by-19 grid board as possible. Individual stones have only a few behaviors, but the large size of the board, the complex patterns stones can form, and the challenges created when trying to join isolated stones into stronger groups while fighting back attacks make the game difficult to master.

Musō Sōseki is a figure roughly equivalent to Christopher Wren, John Donne, and Thomas Becket combined. On Tenryū-ji and Musō Sōseki's designs and influence, see François Berthier, *Reading Zen in the Rocks: The Japanese Dry Landscape Garden,* trans. Graham Parkes (Chicago: University of Chicago Press, 2000), and Katherine Anne Harper, "Daiunzan Ryoanji Sekitei—the Stone Garden of the Mountain Dragon's Resting Temple: Soteriology and the Bodhimandala," *Pacific World,* n.s. 10 (1994): 116–30.

INDEX

ABOUT THE AUTHOR

ALEX SOOJUNG-KIM PANG has spent the past twenty years studying people, technology, and the worlds they make. A professional futurist with a PhD in the history of science, Pang is a former Microsoft Research fellow, a visiting scholar at Stanford and Oxford universities, and a senior consultant at Strategic Business Insights, a Silicon Valley–based think tank. Pang's writings have appeared in *Scientific American, American Scientist,* and the *Los Angeles Times Book Review,* as well as in many academic publications.